# DATA ROOM MANAGEMENT FOR MERGERS AND ACQUISITIONS IN THE OIL AND GAS INDUSTRY

DEVELOPMENTS IN PETROLEUM SCIENCE 69

# DATA ROOM MANAGEMENT FOR MERGERS AND ACQUISITIONS IN THE OIL AND GAS INDUSTRY

**BOB HARRISON**

Director and Co-founder
Soluzioni Idrocarburi s.r.l.
Rome, Italy and London, UK

ELSEVIER

Elsevier
Radarweg 29, PO Box 211, 1000 AE Amsterdam, Netherlands
The Boulevard, Langford Lane, Kidlington, Oxford OX5 1GB, United Kingdom
50 Hampshire Street, 5th Floor, Cambridge, MA 02139, United States



**Library of Congress Cataloging-in-Publication Data**
A catalog record for this book is available from the Library of Congress

**British Library Cataloguing-in-Publication Data**
A catalogue record for this book is available from the British Library

ISBN: 978-0-444-63746-8
ISSN: 0376-7361



*Publisher:* Candice Janco
*Acquisitions Editor:* Amy Shapiro
*Editorial Project Manager:* Danielle Mclean
*Production Project Manager:* Debasish Ghosh
*Cover Designer:* Alan Studholme

Typeset by TNQ Technologies

*The textbook is dedicated to the late Professor Colin Wall, who was Head of Department when the author attended the MSc in Petroleum Engineering at Imperial College London in 1981. The author, who subsequently worked alongside Colin and became friends, fondly remembers the animated discussions over lunch in various public houses in South Kensington in the engaging company of Colin and the late Martin Ford MBE, former Chief Petroleum Engineer of British Gas E&P.*

# Contents

# About the Author

**Bob Harrison** is a petroleum engineer, who has worked in the oil and gas industry since 1979, gaining broad technical and commercial experience around the world, with a strong reservoir engineering and petrophysics background.

After a spell of running wireline logs in the Borneo jungle and shooting seismic in the Libyan Sahara, Bob then worked for over 20 years with independents British Gas E&P, Enterprise Oil, Veba Oil & Gas, and Mærsk Olie og Gas. This was followed by a further two decades counseling many large and small oil and gas companies around the world, both as a freelance advisor and as subsurface technical authority with a global energy consultancy. Bob specializes in due diligence, petroleum resource audits, asset screening, field development plan review, and energy project management. He is an experienced analyst of field performance and reservoir surveillance and has spent a number of years in onshore and offshore well operations including logging, running well completions, and well testing.

He is co-founder and director of Soluzioni Idrocarburi s.r.l., who provides technical consultancy and in-house training courses for the petroleum industry. Bob has an Executive MBA from Cranfield School of Management and received his MSc in Petroleum Engineering from Imperial College London, where he was academic visitor and external lecturer in production logging and petroleum project economics. He also developed and taught the energy project management module at Clausthal University of Technology in Germany.

Bob is a Fellow of the Energy Institute. He has served on the Society of Petroleum Engineers (SPE) Oil & Gas Reserves Committee (OGRC), which maintains the Petroleum Resource Management System (PRMS).

Also, he is past Chairman of the Society of Petroleum Evaluation Engineers (SPEE) European Chapter, responsible for reserve auditor education. Bob is a certified Competent Person, with global authority to sign off resource statements for stock exchanges and government regulators.

# Preface

Whenever companies are involved in mergers and acquisitions (M&A) of oil and gas assets, whether they are exploration ventures, appraisal opportunities, discovered accumulations that are undergoing development, fields that are on production, or mature assets in their declining years, it is almost certain that a data room will be involved during the due diligence process.

Setting up a data room requires as at least as much (and frequently more) forethought and planning as attending one, and there are many elements that are required to be in place for it to be successful. If the attention to detail is lacking, from either the seller or the potential buyers, then the data room is unlikely to achieve its objective and the potential deal may not only be delayed but also it may fail to materialize.

From the perspective of sell-side M&A, vendors must divulge exhaustive information to advisors and potential acquirers, whether they are marketing the opportunity widely or targeting a specific buyer. Sellers either want to raise sufficient capital for company growth and development or to divest certain noncore assets at a healthy price. To achieve these goals, they must attract as many serious potential buyers as possible to the data room, in which appropriate data are provided that are enough to inform and persuade likely investors to make an offer in the scheduled timeframe. Sellers must lay out their most confidential information data for scrutiny by rival companies, investors, and banks. How the data room process is managed, from starting the sale and providing information to negotiating and closing the deal, is a direct reflection of the value of the venture and the vendor. For example, in the case of an online data room, it must be accessible, quick, and easy to use, be a conduit for regular communication through the transaction period, and yet still provide sellers with a sufficient level of control over the security of their important data. There is a natural tendency in data rooms for the seller to accentuate the positive and play down, or even sometimes obscure, the negative aspects of the assets. This can be exacerbated by the cognitive bias that often creeps into business development activity as the fever to "do the deal" takes hold.

Looking at a deal from buy-side M&A, potential acquirers must gather and review huge amounts of data to evaluate a deal thoroughly enough to make an informed bid, but quick enough to contend with rival interested parties. The due diligence of a deal differs from buyer to buyer, but always involves some degree of technical and commercial analyses, which are time-consuming when done meticulously. Buyers want to know four things about any potential asset purchase: volume, uncertainty, value, and risk. Hence, any visit to a data room must be made with the goal of furnishing prospective buyers with an informed view of the range of the estimated recoverable petroleum from initially or currently in-place volumes; what value might be attributable to these volumes, and what risks are associated with their recovery. Shortcuts to asset evaluation such as the use of analogs and rules of thumb that apply estimates of value per unit volume, which may or may not be appropriate, can reduce the decision-making time and so help a buyer to avoid missing investment opportunities. However, these shortcut approaches are riskier and mean the buyers must evaluate deals from a less knowledgeable position. A degree of skepticism is often a valuable ally in the endeavor to provide a full and unbiased evaluation which will facilitate more rigorous decision-making.

This book introduces frameworks and workflows that help to streamline the data room process, to highlight the essential data that must be assembled in the permitted time window, and so accelerate the subsequent assessment of the opportunity. This book format combines theory with case studies, some of which describe lessons learnt through bitter experience by the author himself. Methodologies are presented that can be used immediately by those involved in the technical and economic evaluation of oil and gas exploration and production opportunities. Yet, it is also suitable for readers with a wide spectrum of experience, from those who are newcomers to the strange world of data rooms to those diehards who may have spent too many hours in them.

The purposes, strategies, and tactics of data rooms are explained, along with some suggestions on how to survive them, and how to get a fit-for-purpose evaluation in front of the decision-makers in the shortest timeframe possible.

# Acknowledgments

The author is indebted to fellow oil and gas industry professionals Paul Jeffs, Tim Lines, and Patrick Quinn, who acted as book reviewers and who made many useful suggestions on how to improve the content and the text. Thanks are due also to the author's former colleagues, Lena Ålvang and Patrick Quinn (again), for their help and advice on the geomodeling and dynamic simulation sections of this book, respectively.

# CHAPTER 1

# Due diligence, definitions, and doubt

## Contents



Before one embarks on discovering the world of data rooms, it is essential to recognize that all parties do not necessarily speak the same language. Whenever the term "reserves" is used, all interested parties in a potential asset deal should know what it means and have the same definition in their minds.

One must never lose sight of the fact that M&A activity means that money (and we mean "lots of money") may change hands, so there is a requirement for due diligence in any oil and gas transaction, as investors should never accept anything at face value. Potential buyers must ensure that a seller's resource estimates are "properly classified, accurately calculated, and appropriately reported" (Deloitte, 2013). To achieve this goal, buyers need to be diligent, knowledgeable about resource reporting requirements and limitations, and read carefully the provided memoranda to examine a seller's claims.


*Data Room Management for Mergers and Acquisitions in the Oil and Gas Industry*
ISSN 0376-7361
https://doi.org/10.1016/B978-0-444-63746-8.00001-6

This chapter explains what due diligence is and why it needs a framework that defines how resources should be reported. This section also introduces the main definitions, conventions, and jargon commonly used by M&A professionals in the oil and gas industry. In this way, it is hoped to get all readers "on the same page" whenever the petroleum resource lexicon is used throughout this book. The fundamentals of a petroleum resource reporting system are described here briefly. Finally, the need for analysts to harbor doubts when auditing hydrocarbon resource volumes is raised.

## 1.1 What is due diligence?

In the world of business, due diligence is defined as the procedure of methodically examining and confirming the accuracy of a seller's statements about a deal on offer. Its objective is to ensure that all interested parties associated with a potential deal have the information they require to assess its value and risk accurately and come to an informed business decision. To that end, potential buyers need an independent, neutral, and wide-ranging audit of both the current financial health and future prospects of the assets in the deal. The price of the bid, if one is made, will almost certainly depend on the results of the due diligence, which should verify all material facts concerning the deal, be they technical, commercial, or legal.

Sellers may perform their own due diligence of buyers to confirm that they are neither time wasters nor tire kickers (see Section 5.1), and that they have the necessary financial backing and technical resources to complete the deal in the allocated timeframe.

What one expects from due diligence depends on how the results are to be used. For example, the exercise may be to underpin an asset's sale value, or to provide assurance to a concerned party about certain technical or commercial matters. Alternatively, due diligence may form the basis of a Competent Person's Report (CPR) that will be prepared by an accredited auditor, in an objective and independent manner, for a public listing on the stock market or to submit to investors to accompany and underpin a request for funding. Please note that the CPR is often referred to as the Qualified Person's Report in some parts of the world.

Although the scope of each due diligence exercise is contractually agreed, the evaluation itself is left entirely up to the auditors. Guidelines on the required depth and detail of the work to be performed during due diligence are rarely provided, so the standard can (and does) vary considerably, particularly the estimation and reporting of resource volumes. Due diligence must be thorough enough to permit the buyer to understand the volumetric, recovery, and commercial uncertainty.

There are many factors that affect the integrity of a due diligence exercise: the availability and quality of data, the severity of the time constraints placed on the auditor, and the latter's experience. Regarding the provision of data, the analytical techniques used will depend on the project maturity of the resource being evaluated, i.e., due diligence of a new discovery would need a different approach from that of a mature field with years of production history.

To appreciate the level of work required, one must consider the huge technical effort expended whenever a "raw data to resources" study is performed. The systematic workflow for evaluating any producing oil or gas field begins with raw seismic, geological, and petrophysical data, which are processed and interpreted to construct numerical simulation models to develop an understanding of the reservoir and its behavior. Yet, due diligence is generally done the opposite way, starting with the conclusions, as auditors must review opinions already expressed about the resource and draw on others' understanding of the asset. If the evaluation is restricted to merely reviewing what others have done and relying on what appears "reasonable and sensible" from one's own experience, then such a shallow evaluation should not be considered due diligence. To be classified as such, existing interpretations must be validated at the middle order of the workflow. Spot checks should also be performed at the lower workflow levels to quality control certain elements of the raw and processed data (Hattingh, 2008).

Whichever direction the due diligence takes, from raw data to interpreted resource or vice versa, the buyers' confidence will grow if different approaches give similar results. At the end of the day, successful due diligence comes down to the auditor's experience, and there is no substitute for this.

Ultimately, parties interested in acquiring an oil and gas asset require knowledge about four things: the recoverable hydrocarbon volume estimate and its potential value along with the perceived associated risk and uncertainty.

## 1.2 Assessing volume and uncertainty alongside value and risk

### 1.2.1 Ranges of recoverable petroleum volumes

At the outset of any due diligence exercise there will be a check on the asset's petroleum initially in place (PIIP), which is the total amount of petroleum (which includes nonhydrocarbons) in a naturally formed accumulation or reservoir structure.

To estimate potential recoverable volumes of petroleum, a recovery factor is applied to the estimated PIIP. These technically recoverable volumes (TRV) of petroleum are those that may be produced but are not subject to any economic or legal criteria. This means they cannot be classified as reserves; but more of this is explained later in Section 1.3.

There is a brief reminder of how conventional PIIP estimates are done and what parameters need to be verified during the due diligence exercise given in Appendix 1.

The buyer's multidisciplinary M&A team are tasked with reviewing the provided raw subsurface data, geotechnical studies, legal and commercial contracts, and resource reports to confirm the input parameters of the seller's PIIP and TRV. The inherent uncertainty in these volumetric estimates should be reflected in their low—mid—high ranges and probabilistic distributions, rather than providing a single value.

### 1.2.2 Potential value of produced petroleum

The most popular way of putting a value on a potential acquisition is to build a discounted cash flow (DCF) model of the asset under scrutiny. Within the DCF model, the low—mid—high TRV estimates are expressed as a set of production profiles over time, along with their associated hydrocarbon sales price forecasts (to give revenue streams) and the associated capital and operational expenditures (Capex and Opex, respectively) needed to be invested to deliver those production profiles. The buyer's corporate discount rate, usually the weighted average cost of capital (WACC), in which the debt and equity elements of capital in the buyer's corporate financial structure are proportionately combined, can also be applied. The range of WACC for global E&P companies is around 8 percent to 10 percent. Thus, the DCF approach shows the potential amount of cash that the asset might generate, takes into account any tax implications, and so provides a net present value (NPV) for the asset for the input assumptions.

The due diligence exercise must verify that any field development plans (FDP) and future work programs provided by the seller, which will include well performance forecasts and associated costs, are reasonable and achievable in the given timeframe.

While the DCF approach is appropriate for producing assets and for those appraised discoveries about to be developed, its use for exploration opportunities may be less intuitive. For yet-to-be drilled prospective resources, it may be sufficient to use a correlation of NPV versus TRV, derived from DCF models for fields in the region of interest, to assign likely unrisked values to nearby prospective resources.

The assignment of a value to an asset portfolio based on the NPV per barrels of oil equivalent (BOE), derived from past deals in the region of interest, should be avoided. The use of BOE for reporting resource volumes is permitted by regulatory authorities and stock markets, but it is a flawed method when it is extended to asset valuation as it is based on comparative heat energy values of oil and gas, rather than their true sales values. Thus, reserves disclosures submitted in terms of BOE may cause confusion among investors, which is discussed in Appendix 2.

### 1.2.3 Risks associated with petroleum resource exploration and development

The final step in a buyer's due diligence exercise is to identify any risks pertaining to the asset, including what impact they may have on being drilled and developed, plus suggesting what might be done to mitigate the risks, wherever possible.

In the oil and gas exploration sector, it is acknowledged that explorers commonly overestimate prospect volumes, yet underestimate the likelihood of success.

Studies, using data from United States, British, and Norwegian oil and gas companies, show that the oil and gas industry has traditionally reported overestimates of prospect sizes by around 200 percent when compared to actual PIIP volumes found in successful exploration wells. The main contributor to these inflated volumes is inaccurate gross rock volume or its components (area, thickness, and net pay). Look back analysis found that input ranges used in volumetric calculations have been too narrow, so the true variability in subsurface parameters was not captured in predrill estimates, which has led to prospect volumes falling below the low-side of the PIIP ranges determined prior to drilling (Alexander and Lohr, 1998; Brown et al., 2000; Harper, 2000; Ofstad et al., 2000a,b).

A further challenge when predicting prospective resource volumes is that both oil and gas may be discovered in the same region, but with very different economic values. BP significantly underreported the possible presence of gas in its predrill assessments in the Norwegian Continental Shelf (NCS). This implies a bias toward oil when faced with hydrocarbon phase uncertainty, and reflects its greater value compared to gas, which may have swayed the judgment of exploration managers who were keen to see their prospects drilled (Harper, 2000).

This text is not intended to be a treatise on prospect risk and development risk, but, for background, some risking methods that are popular in the oil and gas sector are briefly covered in Appendix 3.

To progress to the investment phase, sellers must demonstrate that their oil and gas discoveries have overcome the risk of development and that the hydrocarbon volumes up for sale are considered to be potentially commercial. This will depend on how good a job the seller has done in appraising the discoveries, whether the seller's production is within the capability of current extraction technologies or simply whether the seller, who may be the operator or a partner, has access to the export sales market along with the required capital to cover the necessary development expenditure (Devex). However, investors must bear in mind that these "contingent resources" represent the near-term future of many oil and gas companies and should not be totally discounted (as some banks are reluctant to lend funds on such resource classifications).

Once on stream, assets will encounter operational risks, and depending where they are located, they will face differing levels of commercial, political, and environmental risks. All of which need to be identified and quantified (if possible) during the buyer's due diligence exercise.

So now we know all about due diligence, let us consider why there is an absolutely essential requirement to have a resource reporting system.

## 1.3 Systems for reporting petroleum resources

### 1.3.1 Why is a resource definition needed?

It is vital to be consistent when communicating the risks and uncertainties associated with the future production and cash flows from oil and gas assets to the industry at large, to the investment community, and to government regulatory bodies. This message was trumpeted over 40 years ago by the father of modern resource reporting, Vincent Ellis McKelvey, who was Chief Geologist at the United States Geological Survey and its director in the 1970s. In tackling the ambiguity in the reporting of natural resources from the earth's subsurface, he is quoted as saying, "*the quantity of usable resources is not fixed, but changes with progress in science, technology, and exploration and with shifts in economic conditions.*" He went on to explain his succinct definition of different levels of risk and uncertainty associated with mineral resources using a simple graphic known as the McKelvey box (USGS, 1976), which is shown in Fig. 1.1. The box shows how resources, depending on their maturity and level of understanding, can move from the prospective stage, through the contingent classification and so become reserves if they pass the commerciality threshold and overcome developmental risks. There are several classification systems for extracted

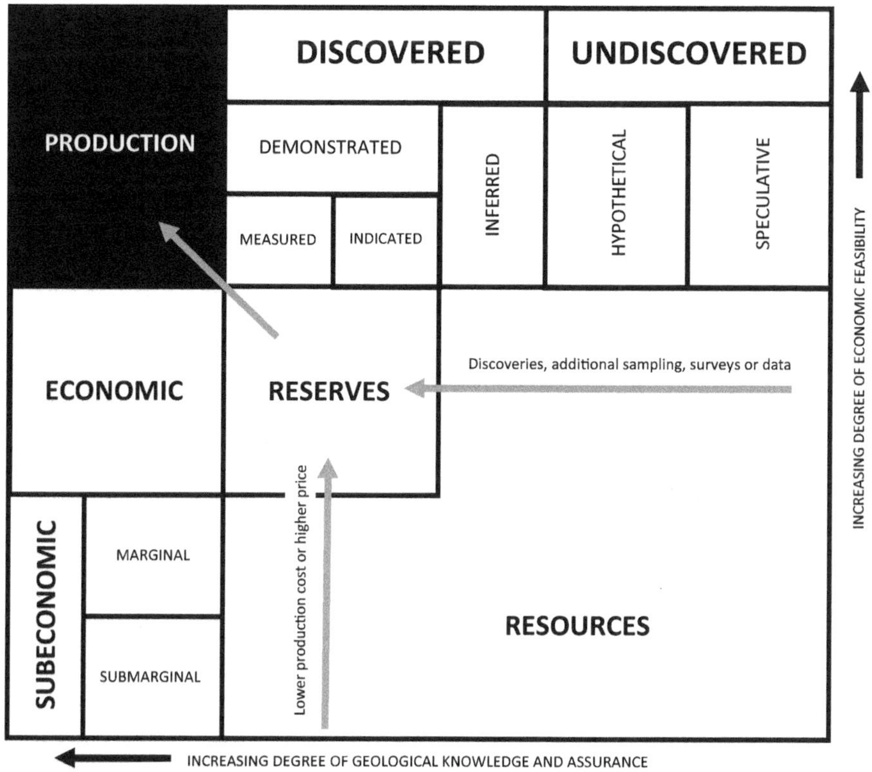


**Figure 1.1** The McKelvey box, updated to reflect commerciality threshold (after USGS, 1976).

resources in use around the world, but they all share the basic tenets laid down by McKelvey. Note that the concepts of "demonstrated, measured, indicated and inferred" relate to the mining industry rather than the petroleum sector.

Today, the industry still faces the challenge of reporting resources reliably and responsibly. Many of the parameters used to define petroleum resources cannot be measured directly, so they must be determined indirectly through geophysical, geological, petrophysical, and reservoir engineering analysis and interpretations. This means that, without exception, the estimates of resource volumes, both in-place and recoverable, have an intrinsic uncertainty. As due diligence must provide verifiable and unbiased results to allow fair comparisons between assets, agreed definitions of resources are required to promote uniformity. These definitions are a standard measurement of the assets under scrutiny, quantifying their associated risk and uncertainty via their class and category, respectively.

Reserves auditors face many pressures when involved in M&A activity. Different parties have vested interests in doing a deal, some of which may not be fully aligned. These parties may try to back the resource auditors into a corner or to pull them over to their side (of the argument). Resource auditors must remain independent in their views and stay within the prescribed guidelines they are asked to use (sometimes it is only the guidelines that will save them) but all reporting systems have some gray areas that allow "interpretation flexibility" by their users. These degrees of freedom, especially in the area of whether to classify an asset as contingent resources or reserves, invariably become sticking points that often end up being resolved through compromise and negotiation; a stressful and unsatisfactory state of affairs for all.

Nearly all oil and gas projects can be characterized robustly by using just four descriptors: volume, uncertainty, value, and risk. The petroleum resources that are to be exploited have an associated risk of whether the project will go ahead or not, and the inherent uncertainty in the range of possible size of the accumulation has a direct impact on the measure of worth of the volumes, which takes into account the development cost and the price at which the petroleum might be sold.

Building on McKelvey's pioneering work, a number of different resource classification systems and reserve definitions have been developed. The headline terms and phrases are often the same, but the specifications and guidelines on how they should be applied can differ. Project-based systems are the most popular as they tie in closely with oil and gas industry workflows such as the decision gate system.

A comparison of the major systems used by stock markets and regulators is given in Appendix 4. However, much of the global oil and gas community, along with some stock markets and regulatory agencies, have adopted the Petroleum Resource Management System (PRMS) of the Society of Petroleum Engineers for resource reporting, and this popular (but flawed) set of guidelines is often referred to throughout this textbook.

## 1.3.2 Definition of reserves

There are five requirements for the claimed hydrocarbon volumes associated with a defined project to be classed as reserves by a company; the volumes must be discovered, they must be recoverable, they must be commercial, and they must be remaining, and the company must have entitlement to them. Putting this definition in the context of a petroleum production profile, as shown in Fig. 1.2, reserves are the summation of the future production, from a given date forward, such as up to the end of

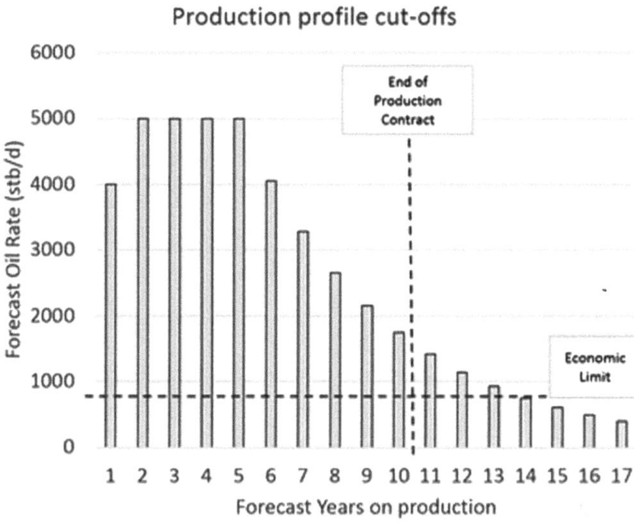


**Figure 1.2** Example of estimating reserves from curtailing an oil production profile by time (license expiry date) and by rate (economic limit is when operating costs exceed revenue).

the production license or when the economic limit is reached, whichever is the sooner. In the case illustrated, the production contract expires in year 10, whereas the forecast net operating cash flow remains positive until year 14, so the auditor must take a view on whether volumes after current contract expiry are reserves. The host government may extend the production license to confer entitlement to the operator, but it may also alter the contractual terms and conditions which could change the economic limit. Alternatively, the host government might award the production license to a rival operator.

One should remember that reserves are a subset of resources that are discovered and commercial, and that proved reserves are the subset of reserves that, in general, are deemed to be over 90 percent certain that at least these volumes will be produced.

As reserves are estimated, there exists a degree of uncertainty in the volumes proposed. A common mistake is to treat all oil and gas reserves and contingent resource volumes as the same and add them together without considering the different risks and uncertainties associated with each of the volumes. As some volumes are perceived to be riskier than others, the reporting of reserves and contingent resources as a single number is frowned upon as it will mislead investors. A probability product model for calculating the chance of development is discussed in Appendix 3.

#### 1.3.2.1 What volumes are not reserves?

With reference to Fig. 1.2, companies should neither claim that the volumes of a subcommercial "tail" of a forecast production profile are reserves nor should any volumes forecast to be produced after the production license has expired be booked. This is the case for rules-based resource reporting systems, but the principles-based reporting systems are somewhat vague on the matter, leaving it up to subjective judgment. The latter gives rise to cases where resource auditors may adjudge it likely that the production license will be extended and so classify these later volumes as reserves but assign them to the higher uncertainty category of either probable or possible. In addition, as it is likely that future production beyond the end of license may have inferior terms and conditions imposed by the host government that will impact on the project economics, auditors may alter their resource estimates accordingly. Yet, these approaches make many unsubstantiated assumptions. Furthermore, they may be applying different contingencies to a defined project's recovery, which is known as split conditions and is not allowed in some resource reporting systems.

Companies often argue that governments traditionally renew production licenses for long-term producers in their countries with whom they have developed a special relationship. Yet, caution is advised when believing what has happened in the past will hold true for the future. For example, in 2017, Qatar Petroleum chose Total SA to operate the country's Al-Shaheen development (which is responsible for over 40 percent of the OPEC nation's output) and replace AP Moller—Maersk A/S in the running of the giant oil field, so ending a 25-year working relationship. Unless producers can show tangible evidence to the contrary, potential investors may be advised to assume that producers will have no contractual entitlement to volumes produced after the production license expiry or they could face an embarrassing write down of reserves in the future (see Section 1.4).

Impurities, such as carbon dioxide ($CO_2$) or nitrogen ($N_2$) in natural gas, do not classify as reserves under the PRMS, unless the gas is sold with the inert constituents included, leading to a lower sale price due to the inferior composition of the gas which would have a correspondingly lower calorific value.

The systems vary on their treatment of produced gas used for fuel and flare. While some state categorically that fuel gas, also known as "consumed in operations," cannot be reported as reserves, others state it can be. Harrison and Quinn (2018) believe it would be misleading to investors if fuel gas volumes were reported as reserves, as they are consumed during production and never reach the market. Instead, they should be treated as a "shrinkage of production," subtracted from produced volumes to provide net sales gas reserves.

The economic benefit of not having to buy gas from external suppliers is captured in the asset DCF model and this will extend the commercial life of the asset and increase its reserves. To book fuel gas as reserves and to claim an economic benefit in lower asset operating expenditure would appear to double counting. Buyers should beware as Ryder Scott (2018) estimates that, on average, 10 percent of the gas reserves quoted by US producers are actually consumed in operations and are never sold on the market.

Generally, royalties that are to be paid in kind from hydrocarbon production are owned by the government, so these volumes are deducted from the licensees' reserves.

And finally, all volumes that have already been produced prior to the effective reporting date of the resources report do not qualify as reserves.

### 1.3.3 Sources of risk and uncertainty

Risk and uncertainty are often used interchangeably, but they are quite different concepts within resource reporting systems. Risk is primarily associated with the classification of volumes and is a measure of the likelihood that the project will be developed. Uncertainty is used to categorize the project's recoverable volumes and is a measure of the impact of technical factors on producibility and recovery.

Sources of risk include the (more likely) chance that the sellers' prospective resources might not be there, but buyers will have to fund an exploration program to find out. There is always the chance that contingent resources may not be developed, especially if the acquisition of new information paints a less flattering view of the project. With regard to influences outside the operator's control, a license may not be extended by the government of the day; or indeed the asset may be expropriated by a regime and/or lost through the courts. The latter is an unfortunate occurrence that will necessitate legal and political due diligence by potential buyers to quantify the risks involved.

With regard to ambiguity in the estimated resource volumes, buyers must answer three questions, "How big is the 'tank'? What can be drained from it? What is the extracted volume worth?" Regarding the size and extent of the resource, the major sources of uncertainty are seismic depth conversion, reservoir continuity and quality, and location of the fluid contacts, where there may be some distance between the lowest known hydrocarbon and the highest known water discovered in the reservoir. The amount that can be extracted from the reservoir is dependent on rock and fluid properties, drive mechanisms, and well productivity. Uncertainty in the value of the recoverable resource volumes stems from factors such as future cost escalation; unclear project definition and execution; schedule slippage; oil and gas price fluctuations; unforeseen changes to the fiscal regime; and, in certain parts of the world, theft of production by criminal gangs.

### 1.3.4 Deterministic and probabilistic terminologies for reserves estimation

Reporting systems permit resource volumes to be calculated from either deterministic or probabilistic methods, yet the estimates produced from each of the approaches rarely agree (Cronquist, 1991).

There are two deterministic approaches to estimating reserves; the incremental method and the scenario (or cumulative) method. The main difference is that the former calculates each of the proved (P1), probable (P2), and possible (P3) volumes separately, whereas the latter estimates low (1P), best (2P), and high (3P) volumes, and then derives the P2 and P3 volumes by subtraction, e.g., probable equates to 2P minus 1P; possible equates to 3P minus 2P. Readers are advised to make note of the number and letter conventions; woe betide any reserve auditors that mix up their 2P volumes with their P2 volumes or their 3P volumes with their P3 volumes. The two deterministic methods should give similar estimates, but often do not, as the principle of "range of uncertainty" is not reflected in the incremental method.

The probabilistic approach uses inputs of estimated ranges and distributions of the necessary geological, petrophysical, and engineering data to generate a spread of reserves estimates and their associated probabilities. The technique often used is known as Monte Carlo simulation, where the volumetric estimation is performed tens of thousands of times to generate a probabilistic range such as that shown in Fig. 1.3. The resultant output curve can then be used with the accepted confidence levels set within resource reporting systems.

Readers should note that resource reporting systems tend not to assign directly any probability values to deterministic estimates, preferring to state they have "broadly inferred confidence levels." Yet, Fig. 1.3 implies there is such a link and many in the oil and gas industry (mistakenly) use the resource categories from deterministic and probabilistic approaches interchangeably. For the record, the implied, but undefined relationships between those estimates of recoverable volumes as calculated from deterministic and probabilistic methods are shown in Fig. 1.4.

Confusion surrounds the two methodologies, especially when and where they may be best applied. The PRMS states that P90—P50—P10 probabilistic estimates should be "reconciled" with low—mid—high

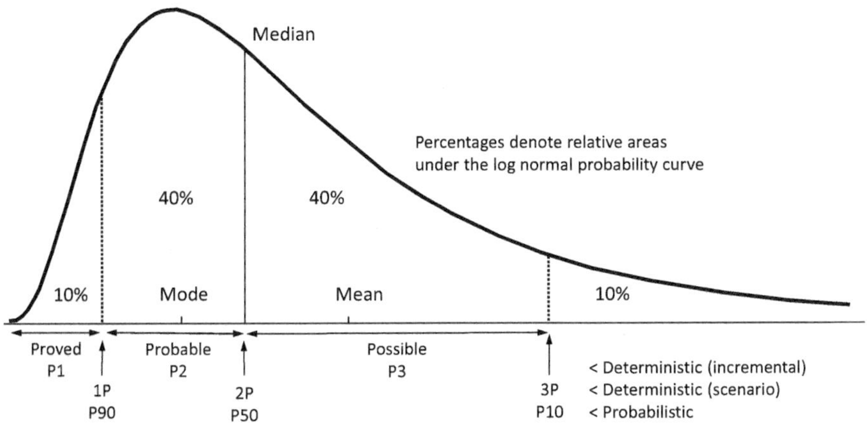


**Figure 1.3** Theoretical probabilistic distribution of reserves, annotated with associated levels of uncertainty for the different reserve categories; incremental/scenario estimates and deterministic/probabilistic estimates often disagree (after Demirmen, 2007).

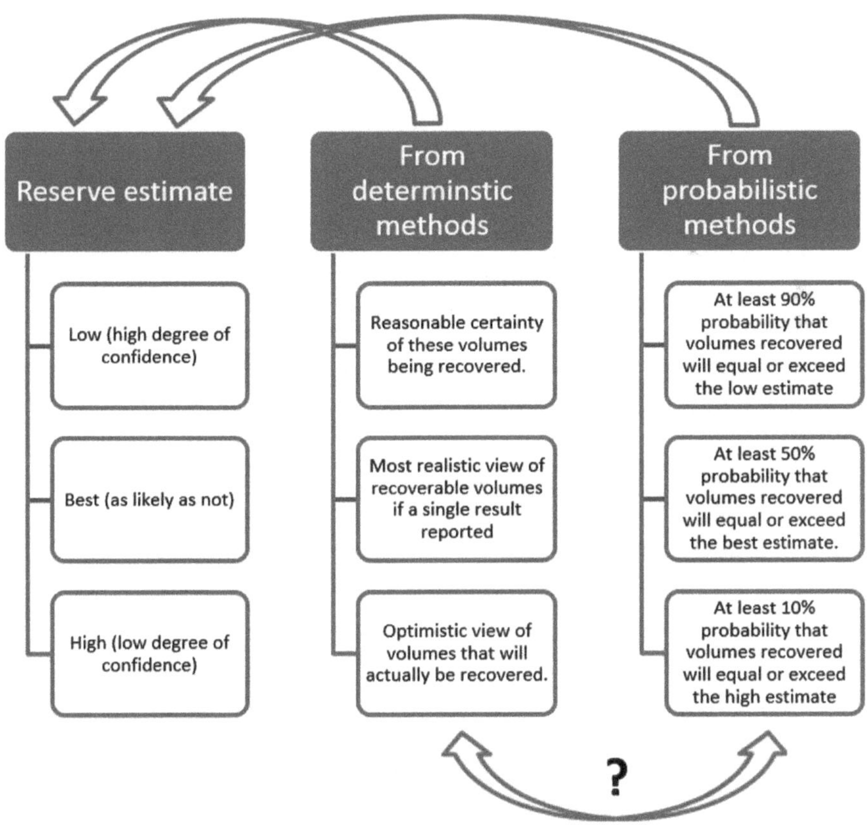


**Figure 1.4** Implied, but undefined relationships between deterministic and probabilistic estimates of recoverable volumes for increasing levels of confidence.

estimates from the deterministic approach, but this guidance is rarely, if ever, followed. Resource estimates should reflect the quantity and quality of data and state of knowledge rather than selecting a different calculation approach with the same dataset.

There is some support in the industry for the multiple scenario method, which is essentially a deterministic method that provides a pseudo probabilistic result. The method describes a range of possible outcomes or subsurface realizations that are physically consistent with the observed data. A single outcome within the estimated PIIP range defines one subsurface realization and the recovery factor is estimated by defining the development scenario for each realization. These scenarios are then translated into a pseudo probability distribution function by assigning chances of occurrence which combines elements of the deterministic approach with the probabilistic method (SPE, 2011). The approach is similar to that used in experimental design, which many operators use to derisk field developments.

The uncertainty in estimated ultimate recovery (EUR) tends to diminish with time and acquisition of additional data; however, oil and gas industry professionals have a tendency to predict narrower ranges than their knowledge warrants (Capen, 1976). A useful reality check on a probabilistic range of recoverable resources is the value of the P10/P90 ratio (Rose, 2001), which is shown graphically in Fig. 1.5. Readers will observe that if assets are under appraised, their inherent uncertainty is higher, and so their range of likely values of PIIP and TRV will be wider. This is an essential check to be done during due diligence.

Having established the spectrum of resource categories and their associated probabilities of occurrence, let us now consider the reliability of reporting systems and how much faith should be placed in them.

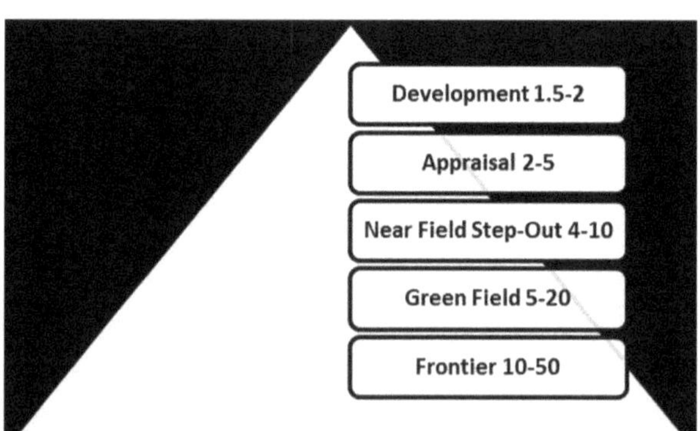


**Figure 1.5** P10/P90 ratio pyramid, showing how the range of uncertainty in recoverable hydrocarbon volumes decreases as more knowledge is gathered (after Rose, 2001).

## 1.4 How reliable is reporting of resources?

The year 2004 was an "annus horribilis" for Shell, the oil Major, and El Paso, the US Independent. Shell was forced to downgrade its booked proved reserves by 20 percent or 3.9 billion BOE; the largest write down of proved reserves in corporate history. El Paso was made to debook around 40 percent or 1.8 trillion cubic feet of its previously reported proven gas reserves.

Shell admitted to overstating its booked reserves in Australia, Nigeria, and Oman. In Australia, reserves were booked on the Gorgon field, despite there being no definite market for the gas, no agreed FDP in place, and no firm commitment to invest in this large offshore liquefied natural gas project. In Nigeria, Shell's reserves estimates did not acknowledge any production license expirations and ignored existing operating conditions. In Oman, there was no agreed FDP on which to base incremental reserves, and reported recoverable volumes were not supported by any identified projects (Olsen et al., 2011). The write down resulted in the resignations of several senior executives, loss of global reputation, and incurred fines of over US$ 150 million from the Securities & Exchange Commission. The Shell announcement in 2004 highlighted the shortcomings of oil and gas Majors self-regulating their reserve reporting, as the practice offers scant transparency into how discoveries are classified and categorized (Oil and Gas Journal, 2004).

In El Paso's case, several of its executives were found guilty of colluding to maximize reported reserves in the Americas. In South Texas, El Paso neither adjusted its booked reserves for the smaller drainage areas required in the low permeability Vicksburg sands nor acknowledged the impact of well interference due to larger drainage areas in the high permeability zones. The forecast recovery from Proved Undeveloped (PUD) well locations from future drilling programs was not adjusted for postdrill EUR, which were based on a recovery of 67 percent from predrill estimates, but which should have been lowered subsequently to 39 percent. Results were also biased by El Paso's use of an outdated study on a single field in the region to justify a minimum well decline rate of 7 percent, when a harsher value of 12.5 percent was more representative. In its coal bed methane projects in the Rocky Mountains, El Paso claimed its wells were draining 80 acres only, when in fact historical bookings showed a figure of 160 acres. Bogus "viable" well locations were created by ignoring current economic operating conditions, and 150 PUD well locations were booked on the flimsy evidence of only three tested well locations and two producing wells.

In the Gulf of Mexico, El Paso neglected to incorporate observed mechanical failures in its producing wells, poorer than expected well performance, and revised geologic interpretations in its booking of reserves. Finally, although gas reserves in the Camamu Basin, Brazil, were reported, there was actually no gas sales agreement in place (Olsen et al., 2011).

The use of qualified and experienced independent external resource auditors, adhering strictly to a recognized resource reporting system, would surely have helped to avoid these errors.

## 1.5 Requirement for doubt and skepticism

The renowned American educator and physicist, Richard Feynman (1985) said, "*There is no harm in doubt and skepticism, for it is through these that new discoveries are made.*" Never have these words rung truer as when applied to M&A activity in the oil and gas industry. The prospective buyer's team gathers and analyzes information from data rooms, strives to capture the uncertainty envelope around a deal, and then presents its findings to the client or its management. The due diligence team can use the language of probability to try and ensure a clear message on the risks versus the rewards of a deal is given to the buyer. However, buyers must always be aware of bias—within their teams, among their experts, and in their investment partners—as its presence may cause irrational decisions to be made and incorrect values to be assigned to assets, leading to unsatisfactory deals. If something is uncertain, then buyers should allow it to be uncertain.

Sellers may find themselves providing their estimates of resource volumes and associated risks to skeptical buyers, who may be predisposed to disbelieve their figures, even though they were given in good faith. Therefore, as there is always uncertainty in any resource estimate, sellers may be tempted to provide buyers with estimates at the upper end of the range, to mitigate somewhat an anticipated discount being applied by potential buyers. This practice can be likened to that followed by those taking out insurance for their household goods, where they are advised to inflate the replacement value of their items as insurance companies will unerringly apply a discount to any future claim. It is the author's experience that there is an inflated value factor that crops up regularly when evaluating large oil and gas company portfolios. It seems likely to stem from the largely unaudited majors applying their own version of PRMS to their portfolios and failing to apply the guidelines as rigorously as independent external resource auditors would.

### 1.5.1 Overcoming bias to ensure data quality

Any asset evaluation requires a forecast to be made of its future performance—and there is always uncertainty about the future—so it is imperative to ensure the quality of one's data analysis to minimize the controllable uncertainty. However, this quality assurance becomes more of a challenge if the data are analyzed by different groups, with different needs, as each group can introduce bias into the eventual forecast.

It is always a potential cause for concern in joint ventures, as partners' goals may not be fully aligned, so that partners with different stakes in the outcome may be prone to motivational bias and have an incentive to misrepresent or at best ignore certain unfavorable information. Like it or not, everyone is prone to biases, even if they are unaware of them.

The six main cognitive biases are listed below (Tversky and Kahneman, 1974), and they have been sorted in terms of those most frequently experienced by the author, who considers overconfidence and use of inappropriate analogs to be the worst offenders in oil and gas asset evaluation.

**i.** *Overconfidence*: this the most common bias encountered. It never ceases to amaze the author how experienced oil and gas professionals, who fully recognize the inherent uncertainty on estimating resource volumes, are so confident (to the point of being defensive) about their own PIIP and TRV estimates.

**ii.** *Representativeness (using inappropriate analogs):* bias stems from the use of flawed logic in causality or inference such as when one fails to consider the available data carefully (usually due to time and resource constraints) or one makes a poor choice of analog that is unrepresentative of the asset under evaluation.

**iii.** *Implicit conditioning*: bias comes from using unstated assumptions in one's analysis, which are often embedded in a favorite spreadsheet that is used from one evaluation to the next.

**iv.** *Anchoring and adjustment*: bias is introduced to evaluations when one works off a "best guess." If it subsequently transpires that this guess is wrong, and it should be changed, there is often a reluctance to do so. Instead, one prefers to "anchor"on the initial estimate and "adjust" inadequately for uncertainty.

**v.** *Confirming evidence*: one exhibits this bias when actively seeking data to confirm one's "inflated" opinion of one's assets. Valid data which do not confirm one's view are simply ignored.

**vi.** *Availability*: bias occurs as it is easy to recall information that is prominent in one's mind. This type of bias is often evident in a data room evaluation where time is of the essence and the team takes short cuts to meet the tight deadline.

The key to obtaining quality analysis of data is to understand that bias exists and try to avoid letting bias creep into one's due diligence. Therefore, buyers must focus on what is important to know, rather than what they actually know. The M&A team must always be prepared to challenge assumptions and methods, and never fail to compare a seller's claims with appropriate facts and real data, wherever and whenever they are available. In any due diligence exercise, one must be explicit about what is known and what is unknown.

### 1.5.2 What is reasonable certainty?

Numbers have only one literal meaning, whereas words have many, so it is better to use probability to reduce ambiguity when communicating the basis for one's judgment to management. The problem is compounded when we use imprecise language every day when discussing whether future events will happen. Terms such as likely, hoped, expect, potentially, probably, and possibly all mean different things to different people.

A seminal paper by McLane et al. (2008) asked a very large number of oil and gas industry professionals to sort a list of 10 words describing the likelihood of an event occurring and assign a corresponding probability to each word. The results are startling, especially as the term "reasonable certainty" was associated with a probability of occurrence of around 75 percent, not 90 percent as stated in the PRMS. This implies that these professionals would attribute greater resource volumes to a "reasonable certainty" case than they would for an estimate of proved reserves.

Fig. 1.6 shows the results of the comprehensive study of the use of qualitative language of probability and the corresponding perceived likelihood of occurrence by Mosteller and Youtz of Harvard (1990). The figure gives the accepted expressions for the P90, P50, and P10 values of probability and compares them with the resource reporting terms of reasonable certainty, probable, and possible, which research would equate to probability values of P75, P70, and P35, respectively.

The author hopes that the oil and gas industry will move toward using descriptive words that better define reserve categories. The phrases "reasonable certainty," "probable," and "possible" have different connotations

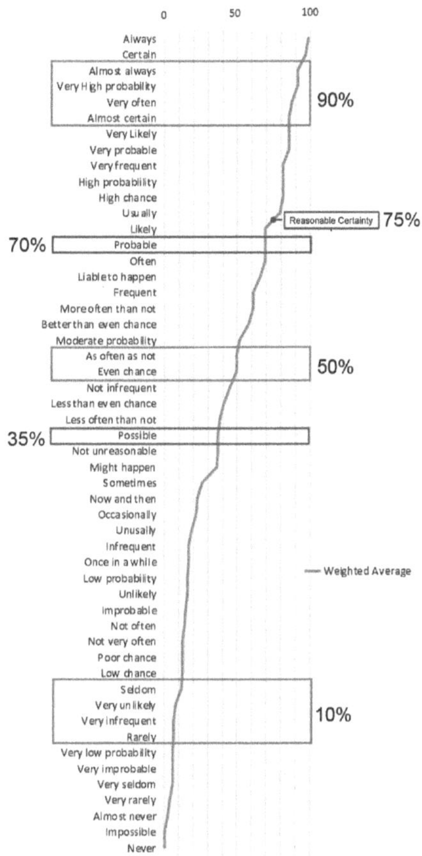


**Figure 1.6** P90–P50–P10 qualitative expressions of likelihood of occurrence from Harvard research compared with language and probabilities used in resource reporting (after Mosteller and Youtz, 1990).

for different people and it would be a tremendous help to investors in the oil and gas sector if these terms were replaced with less ambiguous terms such as "very likely," "as often as not," and "very unlikely," respectively (Harrison and Falcone, 2017).

### 1.5.3 As many answers as there are auditors

The following example (McLane et al., 2008) gives a vivid illustration of just how uncertain hydrocarbon reserves are.

An E&P company contracted with two well-known respected reserve audit firms and asked them to provide deterministic reserve estimates for all of its onshore US producing properties, numbering almost 30 fields. Both

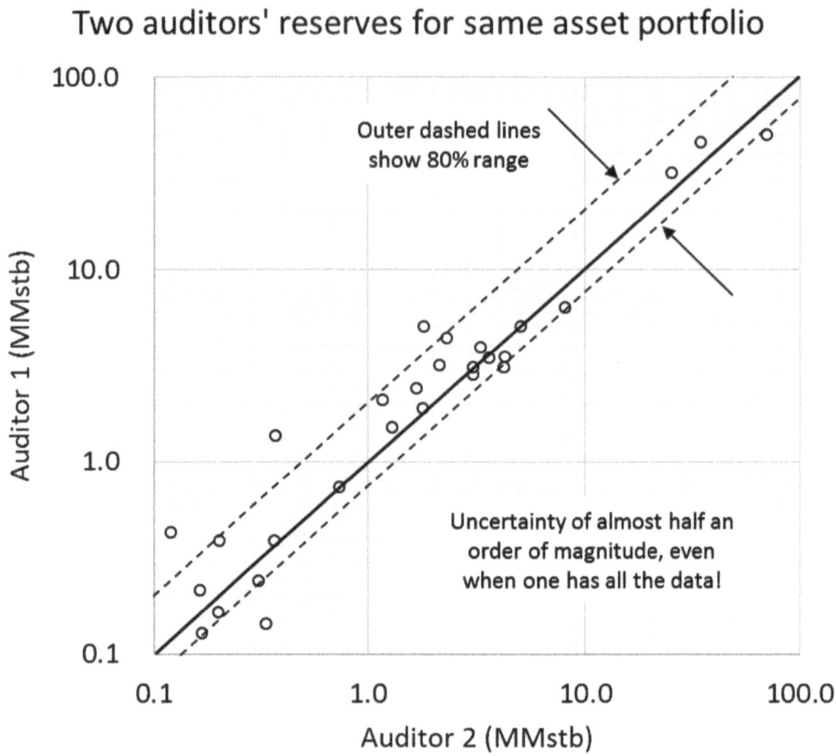


**Figure 1.7** Comparison of reserve estimates made by two independent auditors using the same asset data from onshore US (after McLane et al., 2008).

auditors were provided with exactly the same data. The client then compared the deterministic reserve estimates provided by the two firms on a plot with log–log scales as shown in Fig.1.7.

Auditor 1 booked far more reserves than the second auditor; 22 percent more to be precise, which is a difference of 24 MMstb. This should come as no surprise to those experienced in dealing with subsurface uncertainty. Even when auditors have an abundance of data, as in this case, the inherent uncertainty gave nearly one-half of an order of magnitude of spread in their reserve estimates. This is a clear illustration of the principal drawback of deterministic estimates, which, by their very nature, imply certainty.

Hopefully, readers will now be *au fait* with the absolute need for due diligence during any M&A activity, the requirement for reliable resource definitions, and have ingested a healthy dose of skepticism. Let us now turn our attention to the data room itself.

# CHAPTER 2
# The data room

## Contents



The data room is a potential treasure trove of proprietary intelligence about oil and gas assets that are up for sale, but one must admit that there are more salubrious places in which to ply one's trade.

The author has experienced intense scrutiny by security guards before being permitted to enter a windowless room, with flickering fluorescent lighting and drab monochromatic decor. The seismic workstation and computers sat on bare tables, alongside boxes of musty documentation. The link with the world outside is via the seller's staff, who pop their heads round the door every hour or so to ask if one has any questions or needs any missing reports data to be brought from the archives. This work space is to be one's "home" for the next few days. Welcome to the Orwellian world of pain that is known as the physical data room (PDR).

And the virtual data room (VDR) is not much better, as the M&A team, who has been tasked with due diligence of a large asset portfolio in an impossibly short space of time, can say goodbye to their families and friends for the project duration as they withdraw into their cubicles each day for complete immersion in an online world in order to make sense of a morass of electronic data downloads that is being continually updated by the seller.

This chapter covers the various types of data room one might be faced with and why they are so important.

*Data Room Management for Mergers and Acquisitions in the Oil and Gas Industry*
ISSN 0376-7361
https://doi.org/10.1016/B978-0-444-63746-8.00002-8

## 2.1 What is a data room?

Data rooms are inevitably used whenever oil and gas companies wish to divest or dilute equity in assets or acreage. Potential buyers' teams visit data rooms to inspect, copy, and interpret the data that are deposited there by the seller. The team's findings help the potential buyer to decide whether or not to pursue the opportunity. Even though the materials therein can vary dramatically, in quantity and quality, data rooms remain rich sources of confidential oil and gas field information, but which can be mined only for a limited period of time. These days, data rooms come in different forms and can be physical or virtual or a combination of both.

There is usually no chance to re-enter a data room for a second look once it has been closed by the seller. The data room tends to be a one-off opportunity, so seller and buyers must plan meticulously to ensure it goes well. No matter what the format of the data room is, if the data room visit by the prospective buyer's M&A team, either in person or online, lacks the necessary forethought and administration, it may not achieve its objective and the potential deal may be delayed, or even not go ahead at all.

The following sections describe the different types of data rooms and compare their respective strengths and weaknesses.

### 2.1.1 Physical data room

The PDR has long been a necessary bane of the due diligence process. Traditionally, the PDR is a secure, continually monitored room, either in the seller's offices or those of an intermediary. The author can remember the caffeine-fueled assignments where a buyer's team of experts was assembled to paw through boxes brimming with confidential documents that reportedly held the financial secrets of a prospective acquisition. If additional documents or newer versions of documents are required, these will have to be brought into the PDR by the seller's staff or intermediary. In this "sanctuary," the potential buyer's team will gather for days to comb through reams of confidential data about the seller's assets.

Only one team at a time will be allowed to enter the PDR as rival buyers often do not want their identities disclosed. Teams, often comprised of a number of experts in different fields, may have to be flown in from many regions or countries and remain available throughout the process, so the overall cost of keeping such groups in a data room can be high.

Hence, the PDR is an expensive, cumbersome, time-consuming exercise, entailing extensive data copying, lengthy travel, and tricky logistics. The main disadvantages of the PDR are threefold: location, scheduling, and working in real time. The geographical location of the

PDR may prevent buyers putting in the right team at the right time, and the seller may be unable to update information in real time which may create deal-breaking holes in the asset database.

It should be noted that some governments consider it illegal to take oil and gas field data out of the country. This is indeed currently the case in Malaysia, Indonesia, and some Middle East states. The large oil and gas Majors in the former Soviet Union are also reluctant to let data be taken out of Mother Russia. Hence, the PDR format will always be enforced for M&A activity in these regions.

On a security note, sellers often disable the USB drives in the computers provided in the PDR to prevent the downloading of the data they have provided to be read only. The seller might also ask its office security team to remove the visitor's book from reception, as teams can look back to see which other companies, i.e., potential rivals, have visited the data room before them. Of course, the seller could be devious and insert bogus buyer company names to suggest the deal is particularly interesting and thus more competitive, in the hope of driving up the bid price.

Due to the cumbersome nature and the inherent expense of the traditional PDR, it was inevitable that there would be a move toward an online alternative, and this was realized with the advent of the VDR.

### 2.1.2 Virtual data room

The PDR has largely, but not completely, been replaced by the VDR.

Sometimes called the online data room, the VDR is an extranet or controlled access website that allows thousands of documents to be uploaded for an unlimited number of authorized buyers and their advisers to access and review them simultaneously. Almost all the data released via the VDR are proprietary, so potential buyers must sign a confidentiality agreement to prevent disclosure of any of the material provided to third parties.

The VDR option is arguably as secure as the PDR. It is accessible 24/7 over the allowed period, and it should be easy to administer, and in certain circumstances can be relatively quick to set up. However, the author has experienced up to nearly two weeks of delay in getting through the red tape of various IT departments to gain access to a VDR.

What can take a large amount of time and effort on the seller's part is ensuring that the most appropriate data are gathered and uploaded into the VDR. Information can be added or removed at any time and the changes logged for audit trail purposes. Typically, the cost of a VDR pays for itself in a single M&A transaction.

The VDR option offers many advantages over the traditional PDR method, both to the seller and the buyer. The faster speed, lower cost, and greater efficiency of the VDR allows the seller to access an increased number of bidders and permits a greater bid throughput, as well as providing unlimited time zone access for potential buyers around the world. Experience of M&A activity in the oil and gas sector has shown that buyers get better information, and experience a quicker response time from the seller, which results in more deals being closed successfully. Other benefits of a VDR to buyers include not having to transport their due diligence teams to an out of the way location, which saves time and money, but limits the collection of air miles by M&A team members. Sometimes it can prove difficult for buyers to get the right team into a PDR, but this challenge is overcome with the VDR. Not only can the right team review the deal data but also the information uploaded to the VDR can be refreshed constantly and be immediately accessible. The online system encourages continued daily interaction between seller and buyers via question and answer sessions and requests for new data, which plugs any deal-breaking holes in the evaluation dataset.

A major criticism of a poorly managed VDR is the lack of essential data therein, and the frustrating habit of sellers "drip feeding" data over the duration of the due diligence exercise. This practice delays the progress of the due diligence, causes frustration among the buyers, and extends the time required to confirm the seller's numbers, as often work must be redone when more recent and better information comes to light.

Another common gripe is the temptation for the seller or administrator of the VDR to be overly enthusiastic when implementing controls over access to buyers and their teams. The VDR can provide very high levels of security, including giving individual password access to potential bidders for a given time only, but some sellers can be over zealous in controlling access to their VDRs. The author experienced one extreme case where the seller insisted that all buyers' staff provide names, roles, email addresses, and mobile phone numbers to receive an SMS text each time they wish to access the online data room; once access was granted, the buyers' team had to ensure there was computer mouse activity every five minutes or they were be automatically logged out of the VDR by the security system. One can appreciate the increasing frustration and inordinate length of time required for buyers to download the pertinent data from such a Stalinist VDR arrangement.

Other frustrating security options used by sellers are making files non-downloadable and disabling the screen capture functionality so that users can only read what is displayed on the computer screen while online; or having to submit individual passwords for each file to be downloaded. One appreciates that data security is important, but sellers should remember that the buyers have already signed a legally binding confidentiality agreement to gain entry to the VDR, which means their M&A team members, both staff and associates, are legally bound not to disclose any of the data they are given access to.

Buyers are advised to ensure they designate a VDR downloader for the due diligence team in such instances, as described in Section 2.1.4.

### 2.1.3 VDR and PDR combination

Nowadays, most sellers use a combination of VDR and PDR.

The majority of the documents, reports, maps, images, logs, studies, drilling and production records, etc., are downloadable from a VDR. Yet, some datasets are simply too large to be downloaded in a sensible timeframe. In these less common cases, numbered external hard drives containing the mega datasets are sent to bidders by registered courier to avoid excessive time required to download from the extranet. The large files that are generally associated with processed seismic data and their interpretations, static geo-models, and dynamic simulation models tend to reside on the workstations in a PDR, which is usually located in the seller's offices or those of an intermediary. If the buyer's team members are unfamiliar with particular software installed on the workstation in the PDR, one of the seller's staff or a specialist consultant who are skilled users of the software will be at hand to open and browse the data files and, if permitted under the terms of the PDR, to print off screenshots. These people are only in the PDR to operate the installed software on the workstation, they are not there to offer advice on the data, provide analysis, or comment on the deal to the M&A team.

By combining the VDR approach with that of the PDR, most sellers invite potential buyers to their offices, or at the location of the PDR, to give them their management presentation. This "dog and pony" show allows the sellers to set the stage for the deal and to give their sales pitch to the potential buyers. Sellers gauge the attendance at these management presentations as an indication of just how serious the prospective buyers are in submitting a bid.

### 2.1.4 Virtual Physical data room

In response to the COVID-19 pandemic lockdowns and governments' safety measures, which have disrupted travel and prevented face-to-face business meetings taking place, the ever-resourceful oil and gas sector has developed the virtual physical data room (VPDR). The seller sets up a PDR in a secure location, either the seller's offices or those of an intermediary, and allows potential buyers to remotely access workstations that have been loaded with the seismic data and interpretations along with the corresponding static and dynamic models of the assets on offer. The buyers can neither download nor copy information from the VPDR workstations, but data and models can be interrogated via their own computer screens. During their time in the VPDR, the buyers can interact with the seller and ask for an agreed number of selected screenshots to be saved and emailed to them at the end of the session. The seller's management presentation, which invariably precedes a PDR, may not have the same impact via a video conference as a face-to-face show and tell with prospective buyers, but the VPDR does permit M&A business to continue almost as usual in these unprecedented times. The VPDR is run in combination with a standard VDR as described earlier in this chapter.

### 2.1.5 Designated VDR downloader

To avoid duplication of effort and so save time during the due diligence exercise, a buyer is advised to designate a member of the M&A team to download everything from the VDR and store it on a restricted access area of the buyer's own data server. This approach also ensures that all the data are downloaded, not just those files of particular interest to the certain members of the M&A team. Authorized M&A team members can then quickly access this downloaded version of the entire deal dataset. As updates are often uploaded by the seller, sometimes on a daily basis, it is imperative that the buyer's designated downloader regularly checks the seller's VDR website and keeps the other members of the buyer's team informed whenever new data become available.

## 2.2 Why are data rooms important?

Data rooms are invaluable sources of proprietary information that are, unfortunately, only accessible for a short span of time. However, important and very expensive decisions are made on the basis of a brief spell in the PDR. Thus, the potential buyer's team must gather and analyze all pertinent data to allow the buyer's management to make best decision possible.

There are some less scrupulous companies in the oil and gas sector who indulge in the data room past time of "tire kicking." This unprofessional practice sees the "buyer" claiming to be interested in purchasing an asset, but who is simply using the data room as an opportunity to gain access to proprietary information, often in adjacent acreage to where the "tire kicker" has assets. Sellers must be wary of such companies and can either deny them access to the data room or treat any expressions of interest or potential bids from them with suspicion.

Now let us consider what essential documentation is associated with the data room.

# CHAPTER 3
# Data room documentation

Contents



How do we see what is hidden? Imagine looking at a dog behind a picket fence. You do not see several slices of dog; you see a single dog that is partially hidden by a series of opaque vertical slats. Even though we often cannot see an entire object, the human brain is able to fill in the gaps of missing information and create that image. Scientists refer to this amazing ability as amodal completion (Ramachandran et al., 2010) and it is an extremely difficult thing to do. For instance, researchers have been trying for years, without success, to program a computer to perform the task. So, if our brains are so smart, why do we often struggle to understand and appreciate the complete picture when reviewing opportunities that are outlined in M&A documentation? Maybe it has something to do with the


*Data Room Management for Mergers and Acquisitions in the Oil and Gas Industry*
ISSN 0376-7361
https://doi.org/10.1016/B978-0-444-63746-8.00003-X

pressurized environment in which resource auditors must work and in the way sellers and their advisers draft their M&A documents in legalese.

This chapter describes the trio of M&A documents that turn up in every deal; namely the teaser, the confidentiality agreement (CA), and the all-important information memorandum (IM), from the perspective of the seller and that of the buyer.

By detailing key elements of these documents, the text explains what must go into them and what things are better left out, illustrating how sellers can make their sales pitches stand out from the crowd. These illustrations are also used to give the prospective buyers' point of view and show how potential investors can read between the lines of an IM to formulate searching and critical questions to put to the seller (Harrison, 2014a).

Any buyers or investors who are considering embarking on a due diligence exercise must always keep in mind the phrase, "caveat emptor," which is Latin for "let the buyer beware."

## 3.1 Need for a common language between sellers and buyers

M&A activity is a complex process that is fraught with bias, skepticism, risk, and uncertainty, so effective communication between seller and buyers is crucial to allow the former to showcase the opportunity clearly and for the latter to perform their due diligence to try and uncover any hidden pitfalls. The aim for both parties is to negotiate a fair price for the assets in a scheduled timeframe.

When market conditions are tough, there tends to be a shrinking pool of likely investors, with proportionally increasing demands for assurance from the institutions who provide financial backing for the deal. This will result in extensive due diligence and the lengthening of deal closure timeframes. In these circumstances, sellers must shoulder the burden of knowing and providing precisely what the potential buyers and their backers need to know. If sellers cannot respond quickly to potential buyers' queries and provide satisfactory answers in a timely manner, then they risk either the deal collapsing, or seeing a reduction in its potential value, or having to stumble along a protracted timeline (PwC, 2012; PwC, 2013).

The language of the M&A process can be very different, depending on which party is doing the talking. For example, sellers would like to have a bid price submitted that reflects their arguably optimistic view of the opportunity, whereas the buyers' business development teams, depending on how they are incentivized, might just want to do a deal. Buyers and their backers are seeking the best return for their investment, while recognizing that upside is attractive, but needing to understand its true value and whether that value is commensurate with the perceived risks of the deal.

To this end, whenever oil and gas assets are to be divested, there is a need for a common language between the parties that should be associated with an agreed way of working, so that the different influencing groups can interact effectively with each other and (hopefully) do a deal. Sellers must provide potential investors with their own versions of a teaser, a CA and an IM.

## 3.2 The teaser

At the outset of a divestiture exercise, the seller will usually send out a flyer or teaser to prospective buyers to garner their interest by conveying just enough of what the deal entails.

The teaser is a pared-down IM and can be anything from a single page to half a dozen pages in length. The language used in the teaser is strident and upbeat, in order to whet the appetite of any potential buyers and to encourage them to proceed to the next stage, which is often to register interest in attending a data room.

One can think of the teaser as not being that dissimilar from a movie trailer, which is designed solely to tell the cinema audience that a new block buster is on the way. The trailer gives them a taste of the plot, tells them when it will on their screens, and plants the "I must go and see that" thought in their minds. This is exactly how the teaser operates.

Figs. 3.1 and 3.2 show an example of a teaser, which is advertising a farm-in opportunity in the Southern gas basin of the UK North Sea. Let us go through each of the four pages in some detail.

**i.** The first page of the teaser provides the potential buyer with all the essential information about the opportunity in bullet points. The seller is trying to engage the interest of potential investors by telling the tale of the discovery well, drilled some years ago with an inappropriate drilling mud system that damaged the formation and impaired its ability to flow, so the well rate is not representative of what the undamaged reservoir can actually deliver. The seller's claim may not be unreasonable as the older wells in the region were likely to have been drilled with inferior mud control. Hence, the poor flow test results of the discovery well are explained away in a technically credible fashion by the seller. Whether the seller's claims are true may never be proved (unless another well is drilled into the structure and another flow test performed), but it plants a seed of doubt in the buyers' minds that maybe this opportunity should

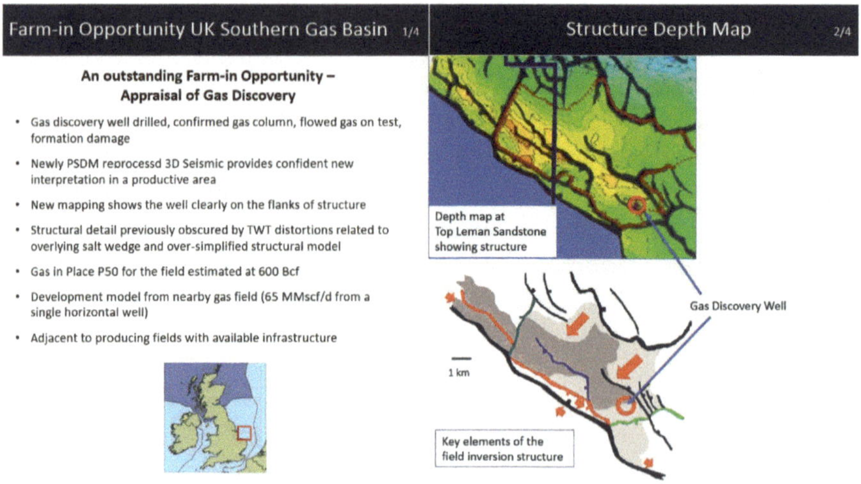


**Figure 3.1** Pages 1 and 2 of a 4-page teaser promoting a farm-in opportunity in the UK North Sea (Harrison, 2014a).

not be dismissed lightly. The promotional spiel continues with a claimed median or P50 value of gas initially in place (GIIP) of 600 Bscf and a proposed field development model based on a subsea tieback to a nearby platform, which is a tried and tested concept used in the region, i.e., drained via single horizontal well. A headline flow rate estimate of 65 MMscf/d is posted in the teaser, along with the seller's

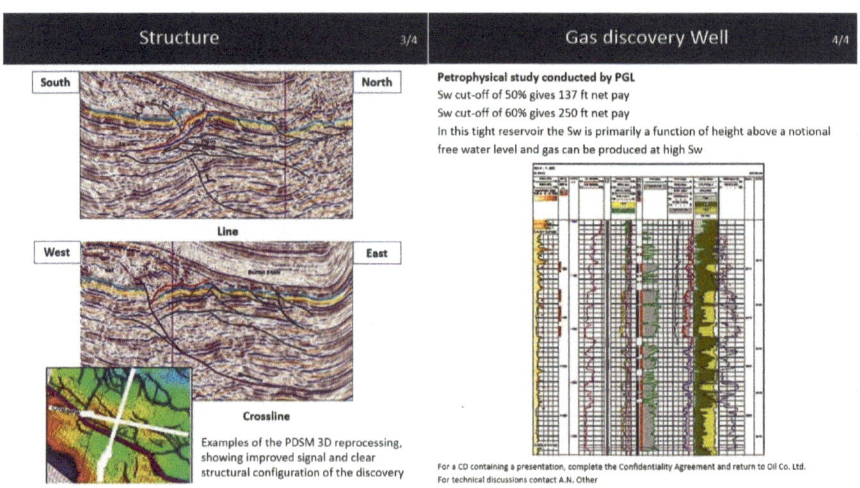


**Figure 3.2** Pages 3 and 4 of a 4-page teaser promoting a farm-in opportunity in the UK North Sea (Harrison, 2014a).

statement that there is access to infrastructure nearby, implying a relatively straightforward tieback to a friendly host platform and a ready export route to market.

**ii.** Colored maps and schematics are used on the second page of the teaser to catch the eye of the buyer. The new map is used to send a strong message to potential buyers that the discovery well was drilled on the flank of the structure, not on the crest, which suggests the new well location will be better placed.
Now let us move onto pages 3 and 4 of the teaser.

**iii.** The geological cross sections printed on the third page of the teaser are used to support the claim that the seller has done some serious geotechnical work, using newly reprocessed prestack depth migration (PSDM) of the 3D seismic data to reduce signal noise and provide a clearer image of the reservoir structure. This image is shown to boost buyers' confidence in the numbers claimed by the seller, who is implying that the new geophysical interpretation has reduced geological risk and that the project economics have been improved.

**iv.** The fourth and final page of the teaser shows a computer processed interpretation of the wireline logs over the reservoir interval. The log data are claimed to have been analyzed by a respected consultancy (PGL). Further, not one, but two values of net pay are given to infer there is relative insensitivity to water saturation cut-off, and so imply that the seller's claimed net pay value is robust. The teaser is rounded things off nicely with instructions to complete and return a signed CA to the seller to get a presentation on the opportunity and gives contact details of the seller and the seller's deal broker for the potential buyers to get in touch if they have any technical questions.

The seller's story seems to hang together and looks plausible enough. However, let us not forget the need for doubt and skepticism that is mentioned in Chapter 1 as we take a closer look at the teaser's content with a more jaundiced perspective.

**a.** The seller quotes the GIIP volume in the teaser rather than the technically recoverable volume. One can surmise that this is probably because the former is always a bigger number, and the fact that recovery may be below average for the play due to the tight nature of the reservoir. The claimed in-place volume of the gas discovery is quite large for a basin that is this mature, and it is likely that the seller has included all the mapped sand in the GIIP estimate, with scant regard for faulting and compartmentalization.

**b.** The headline initial well flow rate is simply that, a headline to grab the attention of potential investors. The stated value is almost double that of horizontal wells reported in this part of the Southern North Sea gas basin, where the main producing reservoirs tend to have poorer permeability and usually require need hydraulic fracture stimulation to produce at commercial flow rates and recover sufficient volumes of gas.

**c.** The seller's claim that there is infrastructure access makes no mention that any buyer will have to negotiate access to the export route, and will have to pay tariffs for processing, compression, and transport.

**d.** Unless the host platform has its own gas compression facility, there will be a severe reduction in recovery from the field, as in the event of a sizable offshore gas discovery, the subsequent development will require a significant pressure boost to reach market onshore.

**e.** There will inevitably be the problem of a new produced gas stream at a higher pressure "backing out" existing production from lower pressure wells of the host platform, which will inevitably incur financial penalties from the host operator.

**f.** The drawing of horizons, faults, thrust lines, and modeled thrust directions on a geological cross section does indeed make the subsurface interpretation look more convincing, even if it may turn out to be incorrect.

And yet, for all the criticisms listed above, this teaser did its job, as the seller farmed out successfully and an appraisal well was duly drilled. Unfortunately, for the investor, the appraisal well confirmed that the P50 GIIP estimate was much smaller than claimed and the test flow rate was far worse than forecast at a measly 1 MMscf/d. The discovery now lies fallow, but the high demand for gas in the region and advances in stimulation technology means there is still hope (albeit faint) that it might be developed in the future.

If potential buyers are intrigued by a teaser, they must contact the seller to request more information about the opportunity, but they would only be granted access to the data if they have signed a CA.

## 3.3 The confidentiality agreement

The CA, sometimes called the Non-Disclosure Agreement or NDA, is a legally binding contract that must be signed between potential buyers and the seller, which defines the confidential data to be shared in order that the buyers can perform due diligence to understand and evaluate the deal on offer. The CA forms a relationship between the parties to protect proprietary asset information being placed in the public domain.

There is a definite buy-side versus sell-side character to the CA; sellers are more concerned with protecting their proprietary information and their employees as broadly as possible, whereas buyers are primarily occupied with limiting the scope of the CA as much as they can.

Buyers are encouraged to ensure that a CA has been checked by their legal department and signed off by senior management before sending their M&A teams to a physical data room (PDR). From bitter experience, the author can confirm that a due diligence team member will not get access to the seller's data room and may spend many nonproductive hours on standby until they have a signed CA in their possession.

The main items that any CA must stipulate are defining what is considered to be confidential information; clearly stating what restrictions are put on parties and who can access the data; the length of time that CA restrictions will last; what will happen if it transpires there has been misuse of the data provided; and what must take place at the end of deal negotiations.

While there is no global standard format for a CA, some guidelines and templates have been proposed (NOGA, 2011; Wright, 2009; DPA/SIPES, 1991). However, any CA should contain some key provisions, and these are listed and explained below.

### 3.3.1 Definition of what is and what is not confidential information

The seller must broadly define the confidential information that will be used in any due diligence. These data may include geological, geophysical, and petrophysical data, maps, models, interpretations, contracts, commercial, and financial data relating to the license area shown on the map in the CA. This map must clearly and narrowly define the boundaries of the geographical area covered by the CA.

Note that the types of information excluded can be critical to protect the potential buyer. Therefore, buyers are advised to exclude information already in their possession, already in the public domain, acquired independently from third parties, or developed independently by the buyers themselves.

### 3.3.2 Caveat emptor disclaimer

Sellers tend to include a caveat emptor disclaimer, which puts the onus on buyers to perform due diligence before making acquisitions, to guard against legal action against them that may arise from providing inaccurate or incomplete confidential information. For example, it behooves sellers to state that they offer no guarantees about the quality, accuracy, or completeness of confidential information provided in the data room. In addition, it is common practice for sellers to ask potential buyers to acknowledge the inherent uncertainty in resource estimates due to the risk of error in the acquisition, processing and interpretation of subsurface data. To round things off, sellers usually go on to state that they have no liability whatsoever regarding the potential buyers' reliance on the confidential information provided.

### 3.3.3 Mental impressions exclusion

The clause about mental impressions can prove to be a little more controversial in that it states while the seller owns the data, the interpretation of the data is the buyer's intellectual property. Thus, potential buyers must insist on the inclusion of a mental impressions exclusion clause that will allow them to keep recollections or memories that they have formed as a result of reviewing the confidential information. The seller must agree that such retained mental impressions shall not restrict the buyer's team from engaging in any aspect of the oil and gas business in regions adjacent to the assets on offer.

### 3.3.4 Return or destruction of data

It is only fair and proper that the buyers must return or destroy all confidential information upon the request of the seller. Of course, the buyers will often claim that any analyses performed by their due diligence teams falls within the remit of the clause regarding mental impressions exclusion and are therefore theirs to keep as they have paid for it to be done.

### 3.3.5 Time limit for CA

Sometimes referred to as the term, a reasonable amount of time should be specified in the CA during which the buyers must keep confidential the information disclosed about the deal. Complications arise if the asset for sale or farm-in has number of partners and other agreements in place. It is common to allow for a period of up to 2 years from the date of the CA, but some seller asks for longer.

### 3.3.6 Breach of CA and choice of law

If a buyer is in breach of the CA, for whatever reason, the seller may (and most probably will) sue for damages. Conversely, if the seller "unfairly" pulls out of a deal, the buyers can ask for their costs to be reimbursed. Of course, it would be a devil of a job for any buyer to prove that a seller pulled out of the deal "unfairly," which is a rather vague term.

In any legal dispute, the seller will invariably insist that the law of its country governs the CA, but if buyers agree, it should allow quicker processing of the case. However, an upfront statement that the English language will be used and that tribunal members can be nominated by the parties will also speed up legal matters.

Having had their lawyers go through the CA and signing it, the buyers can get access to confidential information pertaining to the deal via a virtual data room (VDR), which is in effect a secure web server, and gain entry to a PDR, where a presentation by the seller's team is usually given to the buyer M&A team prior to it gaining access.

At this stage, with the CA signed, buyers will be provided the IM.

## 3.4 The information memorandum

The IM is probably the most important M&A document of the trio discussed in this chapter.

Within the IM, there are the essential and pertinent data required by the potential buyers to make a preliminary assessment of the assets and their value. Like the teaser, the IM must grab the attention of the potential buyers and tell them plainly and precisely about the assets for sale, but in far more detail than given in the flyer. The IM must instill confidence in the likely buyers that the claimed future performance of the seller's portfolio is believable, that the seller's management team are experienced and knowledgeable, and that the claimed value of the asset is credible.

### 3.4.1 The essence of an IM

The IM is not a plan for running the business, rather it is a document to sell the business to professional investors, by giving them all the pertinent data required to make a preliminary assessment of the assets and their value. To this end, the IM has three primary goals:

**i.** *To capture the interest of the investors* which is no small feat as, typically, a professional investment firm can receive numerous IMs every year, so it behooves sellers to make their IMs stand out from the crowd.

**ii.** *Communicate the opportunity to the buyers*clearly and concisely, giving details of assets, infrastructure, operations, and upside. Readers will notice that sellers rarely mention anything about downside in the IM.

**iii.** *Impart confidence* in the business and the seller's management team behind the opportunity.

Thus, the objective of the IM is to establish credibility of the valuation and increase the perceived value of the opportunity in the minds of potential investors.

Consider the opening lines of Bing Crosby's (1941) hit from the Hollywood musical "Here come the waves" in which he sang, *"You gotta accentuate the positive. Eliminate the negative. Latch onto the affirmative. Don't mess with Mister In-Between."* These words could be used by any seller as a blueprint for writing an IM. Herein lies part of the problem with M&A documentation, it must not provide blatant untruths about an opportunity, yet it often contains spin, hyperbole, exaggeration, and over optimism.

Let us consider some of the essential elements of the IM.

### 3.4.2 Reading between the lines of an IM

The typical contents of an IM and some of the more pertinent issues regarding some of the topics are discussed in the following text.

#### *3.4.2.1 Offering procedure*

Before getting into the more contentious material in the IM, readers are urged to note the importance of taking the trouble to read the offering procedure.

It is often regarded as mundane, but the offering procedure, located at the front of the IM, gives times, dates, formats, contact details, addresses, etc., of the due diligence process and how to go about submitting a bid for the assets on offer.

Buyers must not forget to familiarize themselves with these basics, such as making a careful note of the timetable, in order that they know whom to send the bid to, where, in what format and when. It would be a pity for buyers to do perform their due diligence only to discover they had missed the bid submission date or sent it to the wrong person or in the wrong format.

By the same token, the buyer must also check the dates and opening times of the PDR and confirm how many members of the due diligence team will be allowed in at any one time. It would be an expensive mistake to send a large team to a PDR only to discover that some of them are not allowed access and have to remain outside.

#### *3.4.2.2 Importance notice*

This is a statement that should dispel any ambiguity that the sole purpose of the IM is to help prospective buyers to decide whether to proceed with a further investigation of the opportunity, but that the IM is not intended to form the basis of any investment decision.

Wording that is commonly used is, "*This IM does not constitute an offer or invitation for the sale or purchase of securities or any of the business or assets described herein.*"

The disclaimer usually goes on to state any "numbers" in the IM are provided in good faith, but, by their nature, are based on assumptions and are subjective, so buyers should carry out due diligence to satisfy themselves as to the accuracy of all information contained in this IM. Hence, this is effectively a "don't blame me" clause by the seller or the seller's broker.

#### *3.4.2.3 Executive summary*

This summary is a high-level overview of the opportunity, with text specially selected by the seller's team from the main body of the IM to give the most positive spin on the deal. Therefore, it will always put the point of view that the opportunity is too good to miss. Hence, if buyers and analysts are to gain a better understanding of the true pros and cons of the deal, then they must get into the detail of the main report.

#### 3.4.2.4 Confidential information

While not strictly "documentation," data are equally important considerations. Most sellers now use a combination of the PDR and the VDR to provide confidential information to potential buyers (see Chapter 2).

For very large datasets, numbered external hard drives are sent to buyers to avoid excessive downloading from the VDR. Buyers will be expected to formally sign for these drives and return them on completion of their due diligence.

Buyers must ensure that all confidential data provided by the seller are kept safe and sound (and easy to locate), so that they can be returned promptly after the due diligence exercise. It is unprofessional and embarrassing if the buyer cannot find a report, a DVD or external hard drive, etc., and is forced to apologize to the seller for its loss. This unfortunate happenstance can damage the buyer's reputation for future deals.

#### 3.4.2.5 Country overview

This is arguably one of the less important sections of the IM as it tends to report general background information or management overview material. Much of the content often appears to have been lifted straight from articles in world affairs magazines by the seller. Hence, the section is usually not that important for buyers in order to perform their due diligence, so reading of this material can be skipped if one is working to a tight deadline.

#### 3.4.2.6 Corporate organizational structure

The corporate structure of the organization may appear to be straightforward, but it can be one of the more controversial and revealing sections of the IM.

The biographies of the seller's senior management team should be checked to gauge whether they are people one is comfortable placing one's trust in. The potential buyers must decide whether they can work alongside the seller's people and do business with them if the deal involves taking a share in the business as opposed to buying it outright. If it is the former, the investors might be rubbing shoulders with these people for quite some time, so they need to be sure they can get along with them.

Potential buyers should also run a slide rule over the seller's advisors to verify that they have the necessary technical prowess and experience in the region. For example, are they familiar with the seller's assets and the operational challenges that they pose? Do they follow a recognized code of ethics to remain independent and unbiased in their views, regardless of untoward pressure from the seller? Are they accredited by a recognized technical society such as the Energy Institute or the Society of Petroleum Evaluation Engineers? If based in the United States, are the seller's advisers professionally registered and licensed to practice in their State of residence? If the answers to these questions are negative, then the buyers should be prepared to call into question the volumes and values that these advisors have signed off on behalf of the seller.

There is also the adage that "size does matter." If the seller is promoting a venture that is technically challenging, are there sufficient resources and experience within the seller's operations team to deliver the project on time and under budget as claimed in the IM?

### *3.4.2.7 Operational overview*

When it comes to the overview of operations, prospective buyers must never forget that every asset is different and requires scrutiny, and possibly a site visit, to verify that the asset is what is described in the IM. Even within the same country, operating under defined governmental guidelines, it can be immediately apparent that some operators have very different operational practices, safety cultures, and environmental standards. Hence, any buyer should consider a site visit as an essential check on an asset's technical integrity in order to confirm its present condition.

### *3.4.2.8 Reserves summary*

The seller's claims made about reserves in the IM, including the future drilling plans to recover those volumes and the associated costs of those wells, are the meat and drink of the resource evaluator. It is the section of the IM that most auditors head for at the outset of any due diligence exercise.

First off, one should confirm that the provided reserve statements have been prepared by recognized auditors with appropriate experience of the assets and the region.

Buyers should tread warily if it transpires that the seller has a habit of changing auditors regularly as this practice makes tracking reserve movements more difficult.

Another potential cause for concern is when a seller deregisters from one stock exchange to refloat on another with less onerous reporting regulations, such as from the New York Stock Exchange to the Alternate Investments Market (AIM) in the United Kingdom. The latter is an international submarket of the London Stock Exchange that permits small start-ups to float shares with a more relaxed regulatory system than is applicable to main markets. AIM-listed oil and gas companies must report hydrocarbon volumes using the SPE Petroleum Management Reporting System (PRMS), which is a flexible set of guidelines rather than the stricter SEC reporting rules used in the United States.

#### *3.4.2.9 Future drilling schedule*

The drilling schedule provided in the IM is not only used to forecast the future capital expenditure stream but also it provides the basis of future revenues from the realization of undeveloped reserves.

The buyer needs to be comfortable that the seller really does intend to develop the areas classified as reserves that are proven undeveloped (PUD). Some sellers propose drilling schedules in order to get PUD reserves booked by the auditors, but then fail to carry out the development program. In such cases, the seller often claims to have experienced an unforeseen force majeure as the reason why. However, the seller can then go onto claim the same thing the next year, and the next year. This practice becomes more difficult to track if the seller keeps on changing auditors.

Some stock exchanges check how many years the same PUDs have been claimed by sellers. For example, in the United States, the regulator will object if it transpires more than 5 years have elapsed since the PUDs were first claimed.

Buyers need to check that the annual drilling schedule is realistic with respect to availability of drilling rigs, whether the rigs are the right type for the assets to be drilled, and if there is an annual adverse weather window during which no drilling can be done.

#### *3.4.2.10 Lifting costs*

The lifting costs, which are present actual costs and future estimated costs, are essential inputs to the resulting discounted cash flow analysis performed as part of the buyer's due diligence. It is crucial to verify not only what the assets cost to operate today (Opex) but to form some idea of what they are likely to cost in the future. Hence, buyers need to establish the seller's historical operating performance, not only in production and cost management but also to confirm the seller's track record in health, safety, and environment issues.

Buyers must also endeavor to ascertain whether there are any draconian penalties or liabilities waiting in the wings if the operations fail to meet targets and standards set by the country's regulator.

### 3.4.2.11 Decommissioning

One cost that tends to be less well defined is abandonment expenditure (Abex). In some deals, decommissioning is far off in time and so Abex is paid scant attention, but where fields are approaching the sunset of their productive lives, the associated abandonment liabilities cannot be ignored. Governments are waking up to the real challenge of decommissioning and their regulators are demanding ever more information on how current producers are going about financing their future Abex commitments. Whether located onshore or offshore, buyers need to estimate how much it will cost to decommission an asset at its cessation of production, as there may be an enormous impact in the perceived value of a mature asset in the latter stages of its producing life.

Buyers must not forget that PRMS makes an implicit assumption that the funds to cover Abex will be available. In addition, PRMS ignores Abex fund payments, which would be classed as Opex, in its economic limit calculation. Hence, buyers must ensure they have full details of the future Abex requirements of the seller's assets under evaluation (Harrison and Falcone, 2017).

Obviously, the points made above are nonexhaustive, but they hopefully provide a flavor of what potential buyers must bear in mind when faced with an IM.

The IM is the key M&A document, but it is often sprinkled with vague unsupported claims; any perceived upside is trumpeted, yet any downside rarely gets a mention, and project risks and contingencies are often understated. A cynic might wonder whether some of the authors of M&A documentation have studied the Hollywood movie "A man for all seasons," where Sir Thomas More, the brilliant Chancellor of Henry VIII and a Catholic, leaves his Protestant rivals at Court wondering if he had answered their devious questions, mocking them with the phrase, "I trust I make myself obscure."

Finally, it is generally observed that sellers tend to underestimate how much time, effort, and resources are needed to produce good M&A documentation, and some may be better served seeking external professional support from deal brokers when preparing material for an impending data room. However, there are drawbacks to seeking such help in that it can be expensive, it may vary in quality, and sellers tend to have less control over the M&A process.

The next chapter weighs up the pros and cons of using and choosing a deal broker and soliciting financial and legal advisers.

CHAPTER 4

# External help and advice

Contents



The seller can remain in total control of the M&A process by taking a Do It Yourself (DIY) approach to advertising the asset sale, setting up and administering the data room, writing supporting documentation, finding and contacting potential buyers, and managing the negotiations and the bidding procedure.

However, the skills, resources, and time required to achieve these tasks satisfactorily should not be underestimated. Sellers (and buyers) can get support for deal promotion, data room services, and professional advice on financial, legal, and strategic matters, but such external help is not cheap, the specialist services offered can vary widely in scope and in quality, and the seller tends to have less control over the deal.

This chapter discusses the pros and cons of keeping the M&A process in-house or getting external help. It suggests how to go about choosing advisers and what they might cost. Finally, some words of caution are given to those who decide to use deal brokers and advisory services.

## 4.1 Selling the deal

In the quest to grab the attention of potential buyers, sellers normally use a blend of the following courses of action; namely posting the opportunity online, using a deal broker, hiring an investment banking team as an intermediary, or going it alone and promoting the deal themselves.

Buyers who are looking to acquire assets use the same avenues of investigation, but they also have to contend with limited time, as they must

*Data Room Management for Mergers and Acquisitions in the Oil and Gas Industry*
ISSN 0376-7361
https://doi.org/10.1016/B978-0-444-63746-8.00004-1

perform due diligence prior to making an informed offer and do it quickly enough to be able to compete with rival bidders. Unlike sellers, potential buyers must spend enough time to screen and rank the identified deals on offer against their key corporate drivers. These drivers may include the asset's location, its reservoir geology, hydrocarbon type, whether the asset is operated or not, what is its development status, likely price range, and so forth.

Let us look at the different approaches to advertise a deal in turn.

Listing websites contain advertisements of current oil and gas asset for sale. If any of the assets on offer are of interest to buyers, they can pay a subscription fee to obtain a small amount of information about the deal, including notional price and the seller's contact details. Some websites offer to both promote the assets for sale and to auction them online and tend to reveal a little more key data. The obvious benefit to sellers is they get significant exposure of their wares to a large number of potentially interested parties, who can access and contemplate multiple deals. One drawback is that such a blanket approach to the marketing of a seller's assets lacks focus. Potential buyers can trawl through a lot of deals online, but they still need to be screened, with minimal data, to decide whether any of them are an acceptable fit to their corporate strategies. Deals may be missed using this process, depending on what filters are applied in the buyer's web search.

Deal brokers have their own websites on which they list details of asset for sale and they promote deals via online marketing, social media, and by word of mouth to their personal network of potential investors and buyers. Also, they can set up a virtual data room (VDR) and prepare marketing documentation on behalf of the seller. Perhaps unfairly, deal brokers have been described as "commercial real estate agents." In their defense, the more experienced of them have developed a deep understanding the M&A sector and they do know how to pitch a deal to potential investors. The benefit to sellers of using such deal brokers is the creation of a personalized presentation of their assets for sale by the broker to targeted buyers, with organized supporting data to assist the buyers' due diligence.

For large corporate portfolio transactions of high volume and high value assets, about which there tends to be less information in the public domain, sellers often turn to the "secretive" and more expensive investment banking community to act as the middleman to deal directly with identified buyers. Investment bankers help sellers by advising on financial strategy and helping them to prepare the deal. Also, they provide and arrange financing for buyers.

Finally, there is the DIY route taken by sellers who insist on having a major say on the M&A marketing process. The DIY sellers prepare their own material and set up and run their own physical and online data rooms. However, this requires a lot of time and staff resources, and they may not have the niche selling skills that are available via external experts in the field.

## 4.2 DIY or get external help?

Some sellers choose to go the DIY route to retain more control of the divestiture process. In these cases, the sellers opt for a "bare bones" service from an external adviser, where the seller runs the data room from his own offices and writes the information memorandum (IM) himself. However, it is likely that the DIY seller will still have to use the services of an investment bank and a law firm for specialist financial and legal advice, respectively.

There are "one stop shop" external advisers who claim to provide everything, even offering to carry out rudimentary subsurface analysis of the data provided by the seller in order to get the IM out to the investor community in the shortest possible time. This type of service is particularly attractive for small companies with assets to divest, but who either employ few technical people or have had some subsurface work done previously which can be built upon by the broker's team.

Some external advisers work extremely hard once they have been given a remit by the seller, and some claim to be able to launch a sales process within a few weeks if they are pushed. This short fuse delivery is only possible if the seller has all the essential data required for due diligence ready to go. This latter point is crucial.

The author has witnessed chaotic physical data rooms (PDR) which were simply a dump of miscellaneous documents relating to the assets on offer. There is the other extreme where the seller uploads the entire contents of his server regarding the opportunity, including all emails sent and received regarding the assets. Also, there are ill-prepared VDRs where data have been pulled off the seller's staff computer drives, given a cursory quality check, before being uploaded to the data room website. The resulting disorder makes it difficult to confirm whether the right information is being provided in the jumbled, usually out of date, and invariably incomplete datasets to allow adequate due diligence. This always causes delay, so that the IM does not appear until weeks after it was promised to investors.

Sellers should not underestimate just how much time, effort, and resources are needed to set up and run a successful data room and provide good supporting documentation. A solid, instructive, and representative IM can take months to get out to the market.

### 4.2.1 External adviser services

During an asset sale, external advisers should provide some, if not all, of the following services to the seller:

**i.** Help rationalize the seller's asset portfolio and suggest possible buyers.
**ii.** Identify potential assets for divestment and advise the seller on strategy and tactics.
**iii.** Facilitate approaches to third parties and their advisers.
**iv.** Valuation (including analyses of likely competing bids) for the seller.
**v.** Coordinate the divestment process, including assistance with data preparation.
**vi.** Prepare presentations to internal management and the board of the seller.
**vii.** Review submitted bids with the seller.

Of course, the external advisors can also work with potential buyers, and during an asset acquisition should provide some, if not all, of the following services:

**viii.** Seek possible acquisition targets or work jointly with a buyer on a specific opportunity.
**ix.** Screen potential acquisitions and advise the buyer on strategy and tactics.
**x.** Identify potential rival bidders.
**xi.** Evaluate the most likely competing bids.
**xii.** Coordinate the acquisition process.
**xiii.** Prepare presentations to internal management and the board of the buyer.
**xiv.** Help the buyer with his bid submission and subsequent negotiations.

During a divestiture, sellers can simply pay to have the external advisor help to oversee, organize, and execute the sale from start to finish (allegedly). External advisers can also provide guidance on what type of marketing strategy should be employed, i.e., public versus targeted auction, one-to-one sale, or swap, and even identify and target prospective buyers.

The preparation of marketing materials such as the IM, data room setup, and management presentation can all be left to the external adviser, who tends to coordinate the sales process and manage the data room. Brave sellers might even leave the asset evaluation and negotiation of submitted bids to the external adviser.

Similarly, during an acquisition, buyers can ask their advisers to provide their own indicative value of the assets. However, if the advisers receive a bonus if the deal goes through, they may also tend to emphasize the potential upside of the deal at the expense of mentioning the possible downside.

### 4.2.2 Choosing an external adviser

Choosing an external adviser depends on the size of the deal, what elements of the services the seller needs, and whether financing needs to be arranged. Some advisers are keen to provide the financing for deals as well as managing the deal process, seeing themselves as offering the complete solution.

Sellers must read through carefully the external advisers' CVs in order to learn what have they worked on and what experience they have in the country or region where the assets for sale are located. The external advisers' location is important. For example, an adviser based in the United States must contend with a seven hour time difference if the seller is located in Europe.

To remove some of the subjectivity from the selection of an external adviser, sellers can rank advisers and assign marks out of ten for the provision of the services required, weight each service to reflect which ones are relatively more important to the seller and to the deal itself, and then choose the adviser with the best weighted mark.

### 4.2.3 Fees for external help and advice

Obviously, price is a major consideration and, as most external advisers are flexible in terms of what they charge, sellers are advised to haggle. The fee paid to an eternal adviser pretty much depends on what type and level of service the seller wants or needs. Typically, if the assets are sold to someone not introduced by the external adviser, no finder's fee would be payable.

There is nothing to prevent certain external advisers asking for a small percentage of the assets for sale. Note also that the external adviser's contract is likely to specify that the seller cannot easily sway to a different adviser, and the seller may be charged if he cancels the M&A process for unspecified reasons.

External advisers' fees may comprise some or all the following:

**i.** Fixed fee for any technical work performed, which is paid whether or not the assets are sold;

**ii.** Variable fee option 1, based on a percentage of the selling price of the asset(s);

**iii.** Variable fee option 2, structured to provide a large return for an incremental selling price of the asset(s) above a certain level. For example, the seller and external adviser may agree on a target selling price of "X," but the seller will pay out a sizable portion of "Y" if selling price can be made "X + Y" and an even greater portion of "Z" if the selling price achieved is "X + Y + Z."

Deal finders may ask for 0.5 to 1.5 percent of the purchase price, where the variable might be fee based on

**i.** Expected capital expenditure over a given time.
**ii.** Actual capital expenditure over a given time.
**iii.** Over-Riding Royalty Interest (ORRI), based on total production.
**iv.** Net Profit Interest (NPI), i.e., net profit after operational expenditure has been removed.

As a rule of thumb, ORRI is worth up to perhaps 3 to 4 times as much as NPI on a project with high operational expenditure; and particularly if NPI calculation is subject to capital expenditure repayments.

### 4.2.4 Caution with deal finders

Sellers must take care when going out to a large number of external advisers for quotes as this may increase the chance of information leaks. There are always "secret" auctions going on, where the asset sale is not advertised, but the external advisers are instructed to contact "selected parties."

Buyers must guard against deal finders blitzing them by introducing a huge raft of M&A opportunities. Many of these deals may have been unearthed by the buyers themselves, so they must ensure they are not charged a finder's fee for any projects that they were already aware of.

The next chapter deals with preparing for a data room.

# CHAPTER 5

# Before the data room

Contents



Benjamin Franklin, the renowned American polymath of the 18th century, famously said, "By failing to prepare, you are preparing to fail," and nowhere is this advice more apt than in a data room exercise, where every minute counts.

This chapter describes how sellers and buyers should get ready for a data room.

## 5.1 Management responsibilities

From the seller's perspective, its management must decide whether the data room will be a purely in-house affair or whether an external party will be contracted as a "middle man" to administer the process on its behalf.

The data to be provided in the physical data room (PDR), in either hardcopy or files on a workstation, and digital data uploaded to the virtual data room (VDR) must be chosen, located, checked, and made available. It is vital for there to be consistency between the various documentation, data, and models provided. For example, the petroleum initially in place (PIIP) and technically recoverable volumes (TRV) values and associated production profiles quoted in the information memorandum (IM) should come from the static and dynamic models provided in the PDR, which have been matched to the well data and production history provided in the VDR.

To maximize attendance of the data room, the seller should also consider selecting opening dates for the PDR that avoid peak vacation periods, religious holidays and cultural events.


*Data Room Management for Mergers and Acquisitions in the Oil and Gas Industry*
ISSN 0376-7361
https://doi.org/10.1016/B978-0-444-63746-8.00005-3

The seller's management team must also prepare and rehearse its presentation of the opportunity on offer, which will be closely aligned to the contents and message of the IM. This "dog and pony" show is usually presented to buyers at the PDR location, although it could be at the offices of the seller, and occurs before the formal granting of access to the PDR.

From the buyer's perspective, its management must clearly inform the team why they are being sent to the PDR. For example, buyers must decide whether they are sending a team to embark on a "tire kicking" exercise, i.e., use the data room as a data gathering opportunity, or to quality check the seller's existing model to validate the resource volumes and production profiles quoted in the IM, or to create their own fully auditable model of recoverable resources with corresponding production (and injection) profiles.

In addition, the buyer's management should provide its initial ranking of the assets on offer which will be based on the seller's IM, so that those performing the due diligence fully understand what is expected of them. There will only be a limited time to get the data, so if the M&A team members are unsure of what to look for, do not know what questions to ask, or are uncertain why they are there, the PDR visit is likely to result in a poor asset evaluation. There are some simple and effective asset ranking tools available, and some of these are outlined in Chapter 10. One of the most popular tools is the Pareto plot (Barry, 1993), which allows one to rank the assets on offer and to concentrate due diligence efforts on those that contribute 80 percent or more of the asset portfolio's volume or value.

There is often an unrealistic expectation by the buyer's management of the time required to perform a thorough data room exercise. It must be remembered that if more complex goals are set for the M&A team, then this will invariably require more time and more resources. It is unlikely that the M&A team will have a recommendation from its due diligence in front of management on their first day back in the office. Not only must sufficient personnel be allocated to the data room exercise, the M&A team must be allowed enough time (before, during, and after the data room) to perform do a proper due diligence.

Once the buyer's management is clear what assets are on offer, which ones are potentially worth acquiring, what the goals of the due diligence are, then (and only then) should an M&A team be selected for the data room visit.

## 5.2 PDR team selection

Based on the goals of the data room, management must first select the necessary personnel for the M&A team. It is imperative to get the right staff with the appropriate experience and skills into the PDR.

In most data room scenarios, the minimum team would probably consist of a reservoir engineer with some economics expertise and a good understanding of drilling and field production operations; a geophysicist, who will be essential if the seller has loaded some seismic lines onto a workstation in the data room; and a geologist, with some geomodeling skills and a decent grasp of petrophysical analysis. To keep numbers down, one of these M&A team members would be expected to double up as the team leader. However, there are instances when a buyer not only sends some of his own staff to the PDR but also hires an external independent auditor team to attend the PDR on his behalf.

Obviously, the bigger the deal, the more likely it is that the assets offer will be more numerous and diverse, which implies a larger M&A team. The maximum team size would theoretically include one of every discipline, and possibly several of the major disciplines if the asset portfolio on offer is very large. However, it is common practice to pick a fit-for-purpose M&A team for the assets under scrutiny.

In general, if the deal involves relatively more upstream exploration ventures, the team would have more geoscientists, whereas if the deal involves relatively more mature production opportunities, the team would have more engineers.

The interaction between certain disciplines needs to be managed too. For example, geologists and reservoir engineers may be industry colleagues, but they can often be worlds apart on how they approach the data room task. According to Marsh (2009), geologists see the subsurface through rose tinted glasses, they love maps, they are more conceptual and verbose, and they are explorers by nature, who regard failure (i.e., a dry hole) as part of the learning process. Engineers see things in black and white, like graphs, they are more numerate and reserved, they are risk avoiders, and they hold the view that things they create (i.e., a bridge) must not fail. Yet, the M&A team leader must marry together their much needed and diverse skill sets, their attitudes, and their behaviors for the good of the team to successfully tackle the data room challenge.

The PDR workstation software driver (see Section 2.1.4) is often overlooked, yet today sellers often provide full field geomodels and numerical simulation models for review by the M&A team on workstations in the PDR. Note that these static and dynamic models are less likely to be provided in the VDR. If the seller is not allocating someone to operate the workstation(s) provided in the PDR, then the buyer must do so. Therefore, the buyer must ask the seller how many workstations will be provided in the

PDR, what data and models will be loaded, what software will be installed on them, and how many software licenses will be enabled. This information lets the buyer allocate sufficient members of the M&A team who are familiar with the software packages and who will have a functioning workstation to drive. This should be relatively straightforward if the seller is using industry standard software but may be complicated if the sellers are using less popular programs, or even their own bespoke code. For example, some majors build their dynamic reservoir models using their own proprietary simulation software. Thus, an external consultant, who is an ex-employee of that particular seller may be needed to be included in the PDR team.

If the assets have been offered for sale previously, and cases of "it's déjà vu all over again" are more common in M&A activity than one might think, buyers may already have some previous due diligence results on the potential venture. In addition, buyers should ascertain whether any of their staff have evaluated any of assets before or worked on similar fields in the immediate area. Are there any external consultants who have assessed the asset(s) or the area previously that the buyer can call in to help the M&A team? It is an enormous advantage if the buyer can assemble the team with staff that has local knowledge and getting these people onto the team will speed up the evaluation.

The buyer's management must ensure that all team members have a copy of the IM or the data room prospectus or farm-out brochure and that each team member has read it; no-one is exempt.

Finally, it is advisable for the buyer's management to hold a kick-off meeting, to allow the M&A team members to meet face to face, and to encourage any concerns or thoughts to be voiced "out loud" by the team. There is a concern that when buyers send their own staff along with an independent auditor team to the PDR, that the latter's access to key data provided by the seller can be limited by the buyer's staff monopolizing the workstations.

It is usual for the M&A team size to increase as the seller grants the buyer access to the associated VDR. Depending on the assets being evaluated, the buyer would bring in relevant expertise in drilling and completions, facilities and cost estimation, economic modeling, and legal help with contractual reviews.

## 5.3 Travel checklist

All physical data room visits entail many routine, but essential tasks to perform, such as making sure the M&A team has a Confidentiality Agreement (CA) signed by the buyer's management. Buyers should not leave home without one, as their teams will not be allowed access to the data room if the CA is unsigned.

In the post-pandemic age, international travel will take longer due to extra screening of travelers and further virus flare-ups could lead to the reintroduction of quarantine periods when entering a country, which will make attending a PDR very challenging in the new normalcy.

Yet whenever a PDR is elected to be set-up by a seller, depending on its location, there will still be a myriad of arrangements to be made for travel, accommodation, requirement for visas and inoculations, foreign currency advances, or the need for language interpreters to accompany the team, whilst making sure everyone has packed their own personal protective equipment.

In certain countries where a PDR is being hosted, the seller must send individual invitation letters to named members buyers' M&A team to allow them to apply for visas. This process can take some time, so the seller and buyers need to communicate with each other well in advance of the PDR dates to ensure the right people will be in the PDR. Unfortunately, delays happen in the visa application process and sometimes M&A team members requiring visas, even though they may be the best people for the job, have to be left out of the team in order to meet the tight data room schedule.In such instances, a buyer may be forced to hire external consultants that are based in the location of the PDR and have them join his team on site. Besides possible increased cost, the buyer will lose some control over the PDR exercise as the local help are likely to be unknown quantity with regards to their experience and capability.

It is worthwhile, for everyone's safety and security, to make sure that every member of the M&A team has the contact details of the other members of the team, as well as the contact details of the seller's people running the data room.

## 5.4 Data preparation

Prior to the PDR visit, any data that came with the IM should be loaded forthwith onto the buyer's server and the M&A team should be informed where it is stored and how to get access.

Any data provided by the seller that are reported in any language other than English, which is the lingua franca of the energy and financial sectors, should be translated, paper data should be scanned, and log print data should be digitized. Unfortunately, this is often not the case, which adds extra cost and delay for those performing the due diligence.

### 5.4.1 Data uploaded to the virtual data room

In order to allow due diligence to occur, the seller should upload (at least) the following data into the VDR:

- License details and working interesst in the assets on offer.
- Verification that the assets on offer reside within the specified coordinates of the license(s).

- Field Development Plans (FDP), including proposed drilling schedules along with work programs and budgets (WP&B), and documentary evidence of partners' and government approvals.
- Minutes from past meetings of the technical committee (TCM) and the operating committee (OCM) for the assets on offer.
- Subsurface interpretation reports documenting PIIP and applicable recovery factors; low/high gross rock volume (GRV) inputs and assumptions should be made clear and referenced to mapped accumulations. The rationale behind the seller's base case should be explained.
- Petrophysical data and reports, including the minimum log data suite, which should comprise and gamma ray (GR), a porosity log, and deep resistivity. It is also beneficial to provide spontaneous potential (SP), a caliper, and formation tester pressure records. Along with core data, these are usually considered to be the minimum data needed to QC the seller's petrophysical parameters input to the PIIP estimate.
- Drillstem test (DST) data, which should include both surface and downhole measurements of fluid rates, pressures and temperatures; the former by flow testing company and the latter from gauge company. The sequence of flowing and shut-in periods with flow rates are needed to perform an analysis.
- Production histories of oil, gas, condensate, natural gas liquids (NGL), and water, both by well and by field, including uptime.
- Hydrocarbon and formation water fluid properties, from pressure volume temperature (PVT) laboratory reports and assays.
- Shrinkage and yield factors, along with fuel and flare volumes, to convert produced hydrocarbon volumes to sales volumes.
- Geological maps which are tied into static and dynamic models.
- Production and reservoir pressure history versus the operator's forecast.
- Any prior resource estimates, reports, and studies on the assets on offer.
- Decommissioning studies with corresponding abandonment expenditure estimates.
- Post tax economic model, including estimated capital and operational expenditure for planned development activities and associated incremental production forecasts.
- Oil and gas price forecasts, including details of any gas sales contracts.
- Summary of fiscal and contractual terms and conditions associated with the license.

The seller should ensure that the profiles and figures provided in the VDR files match those in the IM and the management presentation to potential buyers.

As VDR data are usually numerous, the downloaded seller's files must be moved to an area of restricted access on the buyer's server. Team members can then quickly access this downloaded version. This stops each member of the M&A team spending hours on the web downloading the same data.

Updates of documentation and data may be continuously being uploaded to the VDR by the seller, so the designated VDR downloader needs to keep checking the seller's designated FTP site and also inform the other members of the team as new data become available.

M&A team members are advised to do some personal research by trawling the web for background material on the selling company. Is there any bad news about the assets on offer? Does the company management and its advisors seem experienced and credible? Does the seller have the necessary financial backing to carry out the work programs claimed in the IM? Are there rumors about whether there are any rival bidders and who they might be? Snippets of news gleaned from such searches may prove useful in the decision-making process.

It is also worthwhile for buyers to check technical society websites for any conference or journal papers that may have been written about the assets themselves, or about offset wells and fields in the area of interest. However, buyers must bear in mind that virtually every conference paper published will tell of a project success story and so have an upside bias; very few articles and papers (if any) will describe a failed venture.

It is usual for the seller to provide a list of what data will be provided in the PDR. Buyers should scan this list immediately, cross-check with what data are in the VDR to identify any data gaps that may hinder due diligence, and contact the seller before the PDR visit with a request for any "missing data" to be made available in the PDR and in their preferred formats to speed up their asset evaluation.

### 5.4.2 Data formats

It is almost always the case that M&A information is provided by sellers or their agents in both paper and electronic forms; although in today's digital world, the preponderance is for the latter.

However, one should not be surprised when working in less developed oil and gas provinces, or when presented with very old fields, that data are often only available in paper form. When this is the case, try and get hold of original copies or at least good quality photocopies of the most important documents. Note that buyers will most likely have to scan and return any original documents to the seller. While this is not an issue usually, it will mean more time is required to get data to the M&A team members in formats that they can review effectively.

If at all possible, the M&A team should try to have data formats pre-arranged and agreed with the seller or the company being hired to manage the data room on behalf of the seller.

An oft overlooked format is that of language. The majority of data rooms are run entirely in English and the data therein are also in English, but this may not be the case when working in regions such as North Africa, Latin America, or Russia and the republics of the former Soviet Union, where French, Spanish, and Russian tend to dominate, respectively. The language issue is difficult to resolve, but it helps to assign one or two speakers of the appropriate language required to the M&A team. While many companies have a smattering of staff who can speak French or Spanish, relatively few have Russian speakers, who are familiar with the Cyrillic alphabet. The buyer's management must be aware that the need to translate data room documentation can add significant cost and time to the due diligence (see Section 9.1).

Raw seismic data are rarely provided in data rooms. Rather the processed data and the seller's subsequent interpretations are installed on workstations in the PDR.

Basic well log data, if provided, should be in the log ASCII standard (LAS) format, but earlier formats such as LIS or the more generic ASCII formats will suffice, as these can be read directly into the log interpretation software packages.

However, the buyers' team should make sure to ask the seller that when the IM states that "log data will be provided in electronic format," it does not mean the pictures of the well logs will be provided as PDFs. Of course, log prints can be traced or scanned for digitizing later, but this is another extra cost and will take more time. Also, having to digitize raw data means that there is always the risk of errors being introduced into the files for review.

Well test data from downhole pressure gauges (i.e., time-pressure-rate or TPR files) may be provided as an interpretation file in commercial transient pressure analysis software. This is also the case for production data, which is often supplied in ASCII, spreadsheet, or commercial database formats.

## 5.5 Equipment for physical data rooms

In these "digital days," almost everything needed to carry out due diligence can be downloaded from the VDR for offsite analysis. However, as mentioned in Section 2.1.1, there are places in the world where it is illegal to take data out of the country, so one's attendance of the PDR is compulsory.

In these instances, as in the "non-digital" past, the M&A team members' briefcases might hold rulers, colored pencils, tracing paper, squared graph paper, log interpretation chart books, phrase books, and foreign language dictionaries. Among the data room paraphernalia, there might also be a planimeter, which is an odd-looking Heath Robinson device for measuring areas contained within the contours of reservoir structure maps.

However, buyers should always check in advance with the seller what equipment will be allowed to be taken into the PDR. For example, some countries will not permit laptops, mobile phones, or digital cameras to be taken into the PDR.

Yet, the author has painful memories that suggest the most important piece of equipment one should take along to a stressful and tiring PDR is probably a bottle of aspirin (Harrison, 2013).

# CHAPTER 6

# In the data room

## Contents




*Data Room Management for Mergers and Acquisitions in the Oil and Gas Industry*
ISSN 0376-7361
https://doi.org/10.1016/B978-0-444-63746-8.00006-5

With a mixture of curiosity and trepidation, the M&A team arrives at the offices hosting the physical data room (PDR). After sitting through a carefully orchestrated sales pitch by the seller's management, they may be asked to show the signed confidentiality agreement to the security personnel before they can gain entry to the PDR itself. The buyer's team hope that the PDR will be an Aladdin's cave of data.

This chapter outlines what the M&A team must achieve during its limited period in the PDR and, equally as important, what it must refrain from doing in order not to waste time. The data room framework and hierarchy are introduced, teams are encouraged to be follow in order that no stone is left unturned, essential data are captured, and imperative questions are asked (Harrison, 2012).

The buyer's team's efforts must be geared to understanding the four essential things about any potential asset purchase: volume, uncertainty, value, and risk. Thus, the aim of the PDR visit is to furnish the prospective buyers with an independent view of the uncertainty in the seller's petroleum initially in-place (PIIP) and claimed petroleum volumes that are likely to recoverable, in what timeframe, at what level of capital expenditure, an estimate of their value, and what contingencies may impact their recovery.

To this end, the roles of the various members of the multidisciplined M&A team are described in detail, including checklists of which data are considered vital and which are nice to have while in the PDR. In contrast, the role of the seller is also discussed.

## 6.1 Role of the seller in the data room

Once an expression of interest in the assets for sale has been made, and the potential buyer has signed the confidentiality agreement, the seller or an agent acting on behalf of the seller will invite the buyer's M&A team to the data room.

The seller has completed the lion's share of his work during the arduous preparation of the PDR and VDR, so his role is to make the subsequent due diligence process run as smoothly as possible.

The seller's management team usually likes to meet the buyer's team and make a presentation about the assets on offer. They invariably put a positive spin on the sale and may use the talk to steer the buyer toward the upside of the portfolio, while glossing over (or neglecting to mention) any downside. The data room time schedule will be clearly laid out, as will the entire M&A process, with particular emphasis on when, where, and in what format prospective buyers must make their formal offers, should they decide to do so.

The seller or agent will act as steward for the data room; ensuring schedules are kept, but also to gather any questions the buyers may have, and deal with any requests for additional data. The buyers' questions are usually asked to be submitted formally to the seller, who registers them and passes them onto its staff who can answer the question or find and provide the additional data. Some sellers share the questions and answers with the other buyers' teams (the buyer's identity being kept secret), other sellers only provide an answer to those who asked the question. In the author's view, the former response by the seller is preferred. It maintains a level playing field during the data room and it serves to advertise to buyers that they are in a competition, which will only help drive up the eventual bid price for the assets on offer.

Occasionally, the seller will sometimes provide access to the technical staff who are presently working on the assets on offer. Buyers should make as much use of them as possible as it can prove to be far more instructive talking to those with experience of the assets than wading through mountains of data.

## 6.2 What to do in the data room

The M&A team must decide on who is going to get what data. The initial allocation of effort tends to be based on the ranking of the assets on offer by the buyer's management, so on entering the PDR, the M&A team will try to access as much as possible of the key data for the perceived "hot assets."

Once in the PDR, it may become painfully clear to the M&A team that the preliminary ranking of the assets on offer, provided by the buyer, was flawed. Hence, it is always useful to have the team reconvene after a few hours of data gathering activity to discuss the data and any preliminary findings. Coffee breaks or lunch often provide a natural pause in the data hoovering process. Not only is this process of repeated review and refocusing of the team's efforts important when dealing with a large number of assets on offer but is also useful when looking at a single field or license block in order to concentrate the team's efforts on key wells and issues.

At the end of each day of the data room exercise, it is proposed that the team holds an informal meeting, usually just prior to be being ejected from the PDR, to assess progress and to gauge overall sentiment about the deal in general. These meetings invariably become an ongoing review and high grading of the assets on offer and, in the event of its daily findings, a new asset ranking can be quickly drawn up, with its associated work program for the remainder of the time in the PDR. These team get-togethers are sometimes dispensed with in favor of grabbing yet more data, yet they can represent a far better use of half an hour.

On a softer issue, if the potential deal is a merger or an acquisition that will include some or all the seller's staff as part of the "package," then the buyer must get an appreciation of the seller's corporate culture. Failing to successfully merge buyer/seller cultures can lead to lingering operational problems, so the buyer needs to understand the seller's norms, beliefs, and values (LaPlante, 2006).

If the PDR is located in the seller's offices, a simple, but effective way of getting a feel for the seller's corporate culture is (if permitted) to go "walk-about" and informally chat to one's opposite number. It is amazing how fruitful this can be, if only to establish a mutually respectful professional relationship.

## 6.3 What not to do in a data room

It is imperative that M&A teams do not dispense with their end of the day team meetings and reviews to grab some more data. Regular communication throughout the team is essential for the PDR exercise to run smoothly and effectively.

Team members should avoid spending hours on end typing data into a spreadsheet. The author remembers a reservoir engineer colleague, who was very keen on pressure transient analysis, spending an entire morning in the PDR typing in time, pressure, and rate (TPR) data from various well test reports. Eventually, the team leader spotted this wasteful activity and encouraged the "typist" to gather more essential data. Of course, there may be some instances where a particular well test is identified to be a key element of the asset evaluation, but even then, the author would recommend requesting that TPR dataset to be uploaded into the virtual data room (VDR) for subsequent pressure transient analysis.

If there are numerous hardcopy reports or scanned data to be captured in the PDR, and the seller has agreed that they can be copied, then it is a much better use of the data room professional's time to take along an external service company hand to type up or scan the data for the team.

Above all, when faced with mountains of information and very little time to review it, the M&A team must remember the important advice on the cover of the Hitchhiker's Guide to the Galaxy and "don't panic." In addition, the team leader must ensure the team do not lose sight of the end game and keep them ticking along in the PDR. It is impossible to remember everything, so the team must not forget to take copious notes of their findings and avoid getting bogged down in minutiae.

## 6.4 Hierarchy of data requirements

As part of the proposed framework for managing data rooms, a suggested hierarchy of data capture is given in Table 6.1 which is intended to ensure that most, if not all, vital information is collected during the usually fraught and frenetic data gathering exercise. These data include technical, commercial, legal, and corporate material.

It is hoped that the structured approach will also help if the M&A team does not have a representative of each discipline. The approach provides an overview of issues and concerns pertaining to each role and outlines what is expected from each during a data room exercise. Hence, team members can ask relevant and searching questions on behalf of technical staff who were unable to attend the PDR. The framework includes checklists of which data are vital, which are nice to have, and which are superfluous background materials (Harrison, 2012).

## 6.5 Role of team leader

In an effort to keep costs in check and to optimize the use of scarce resources, it is best to keep the size of the data room M&A team to a minimum. Hence, one of the disciplines, usually someone at a principal technical level, should take on the mantle of team leader as well as performing their own technical function.

**Table 6.1** Hierarchy of data requirements from a data room.

| Priority | Relative importance | Comment |
|---|---|---|
| 1 | **Essential** | **Absolute minimum of information needed to complete any due diligence** |
| 2 | Valuable | Nice-to-have information that can reduce uncertainty in key parameters, or helps to accelerate the due diligence |
| *3* | *Beneficial* | *Background, nonessential information that can help frame and put into context the due diligence* |

A major part of the team leader's role is to ensure that the M&A team is given sufficient time (before, during, and after) the data room to deliver a quality evaluation report to the buyer's management. The time required will vary depending on the objectives of the buyer and the complexity of the asset portfolio being evaluated. Of course, the deadline is set by the bid submission time and date as given in the information memorandum (IM), but if it transpires that more due diligence is required than the time allows, then more staff will have to be drafted into the M&A team. For example, is a straightforward review and quality check of the seller's case all that is needed or does the buyer require a complete independently built model with associated risk, cost and production profiles? The team leader must always anticipate the need to ask for an extended timeframe (if the seller will allow) or the requirement to increase the size of the M&A team, as invariably the due diligence process tends to reveal unexpected concerns and raise new questions. It is imperative that the team leader ensures that all mandatory data are collected and hold the first review on the first day of data room.

M&A teams must remember to be questioning and skeptical when entering the world of data rooms. They must never lose sight of the fact that sellers will always emphasize the upside and skip over the downside of an opportunity. For a deal to be condoned, the M&A team must be able to convince its management that the necessary information from the data room has been captured, that the downside is sufficiently understood, and that it can be handled.

The following sections outline what is expected from each discipline in the data room exercise.

## 6.6 Role of the driller

Bearing in mind that usually over half of the capital expenditure of oil and gas projects is drilling related, it is odd that having drillers in a data room is a rare occurrence. This may not be the case if the opportunity concerns unconventional resources as these assets are primarily driven the efficiency and cost of drilling and completion (D&C) operations which are often termed "factory drilling" as opposed to subsurface modeling. Drillers are required to formulate the best demonstrated practice (BDP) in the area of interest of drilling and completing a well, as cheaply and as safely as possible. They may need to provide input on the formation evaluation programs in future wildcats as this knowledge will impact on well costs and how the subsurface uncertainty may be reduced. Yet ultimately, drillers are required to verify D&C schedules and estimate corresponding drilling expenditure (Drillex) and for the economic model. Such drillex estimates vary in quality depending on where one is in the life cycle of the project, but they tend to converge and become less uncertain as the project matures as shown in Table 6.2. The driller can develop a better insight into the seller's drillex estimates by breaking them down into their tangible and intangible components. Tangible drilling costs (TDC), which are tax deductible, include the measurable costs of equipment with resale value. Intangible drilling costs (IDC) are incurred by an operator at the well site, but which do not themselves produce a physical asset, such as site preparation, labor and rig rental, which have no salvage value.

Much of the driller's analysis will involve the interrogation of the drill depth versus time and cost curves (Fig. 6.1) for past wells drilled in the area of interest. Drillex tends to increase exponentially with depth, so when curve-fitting drilling cost data, it is often convenient to assume a power law relationship between total well cost, C, and total drilled depth, D, where $C = ae^{bD}$ and the constants $a$ and $b$ will depend primarily on the well location.

The drill depth versus time and cost curves can also be plotted against the encountered geological stratigraphic column and casing shoe depths, and the different time and cost per unit depth calculated for each major formation to be drilled through. This allows a probabilistic model of drilling performance to be developed for the area of interest. It is possible to compute the BDP by adding each of the shortest drill times for each formation penetrated on the way to the reservoir from the wells' time–depth curves and so create a theoretical "fastest well" model for the region of interest.

Drillers need to understand the issues around well planning in the assets under investigation; what problems are associated with the typical formations drilled, what is the availability and quality of offset well records, and what casing programs are typically employed. Other aspects that need to be covered are required permits and certifications for drilling operations in the region and how long it takes to get them from the local authorities. Also, what quality are the rigs that are available, what are the pressure

**Table 6.2** Uncertainty in drillex estimates versus stage of project.

| | **Drilling expenditure certainty →** | | | |
|---|---|---|---|---|
| **Quality class of estimate** | **Order of magnitude** | **Definition of project** | **Exploration AFE** | **Development AFE** |
| 1 | +50%/−50% | | | |
| 2 | | +25%/ −25% | | |
| 3 | | | +20%/−20% | |
| 4 | | | | +10%/−10% |

After Beliveau (2002).

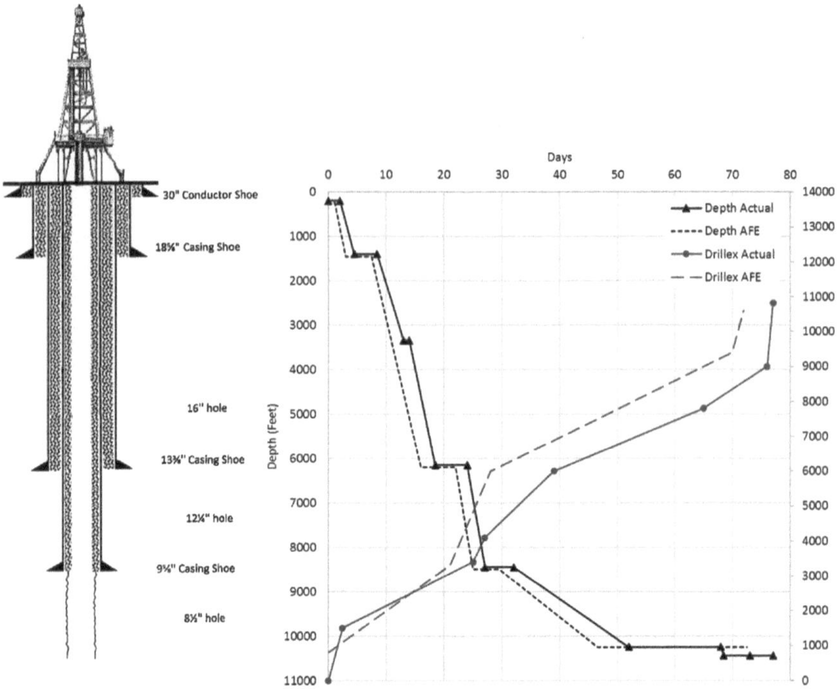


**Figure 6.1** Example of a drilling depth versus time and cost curve, with planned (authorized for expenditure) versus actual performance, onshore well, Middle East, 2017.

ratings of the blowout preventers (BOP) in service, and what environmental considerations and oil spill response capability are needed? The latter items have become very sensitive issues since the Macondo disaster in the Gulf of Mexico (GOM) in April 2010 and the current move into the relatively untouched Arctic frontier.

Let us consider the elements of drilling operations in more detail.

## 6.6.1 Well planning

The M&A team driller must become *au fait* with any drilling challenges associated with the typical formations for the region. These may include abnormal pore pressure regimes with the associated risk of blowout, loss zones that can siphon off drilling mud, reactive shales that can swell or washout and reduce or increase borehole size, respectively, and $H_2S$ bearing zones. Any instances of stuck pipe should be recorded, such as differential sticking (where the overpressure from drilling mud in the wellbore far exceeds the formation pressure); key seating (where drilling or the dragging action of pipe on a dogleg causes a groove to be cut in the side of the borehole, parallel to the axis of the well); and "ledging" (where the drill string hangs up on ledges caused by washouts of softer formations interspaced between harder layers). All the aforementioned factors may contribute to geo-steering directional difficulties. Salt movement is a risk in some areas of the world, which would require the deployment of thicker and stronger heavy-walled casing.

To this end, local drilling practices need to be investigated, including drilling mud type and makeup, whether drill bits of the traditional roller cone type are preferred to those of the polycrystalline diamond cutter variety, and what drilling parameters are employed in the region. Typical casing programs should be reviewed to record, design criteria such as local formation pressure gradients, shoe depths, burst/collapse pressures for casing, and the adequacy of cementing programs. Finally, a look at leak off pressure testing and any recorded casing failures will give an idea how successful drilling has been in the assets.

The location of well sites is another critical factor to be investigated by the driller. Besides the wells' design and construction, are there any issues concerning the terrain itself, such as does the wellsite's condition vary depending on weather and the time of year? For example, the rainy season in the oil-rich Llanos Basin, Colombia, occurs between April to November and causes river levels to rise, making supplying rigs difficult, so that drilling operations are usually suspended until the dry weather returns. Similarly, drilling operations are suspended in the Atlantic Ocean, west of Ireland, as sea conditions from October to March are too rough to supply the rigs. Hence, wellsite accessibility is paramount as this will affect the subsequent production forecasts. In onshore drilling operations, the conditions of the roads and any bridge weight or height restrictions must be scrutinized. Offshore drilling operations throw up different questions, such as whether the well is located close to a busy shipping lane, the possibility of boulders on seabed upsetting jack-up rig deployment, and the likelihood of a blowout hazard due to encountering shallow gas.

Each global region has its own foibles related to drilling planning. To fine tune the drilling time estimate, the M&A team driller needs to ascertain how much red tape will be encountered for drilling permits and rig certification, and how long it takes to get these from the local authorities. Also, how is the region of the assets for sale served by drilling contractors and service companies, who supply drilling fluids, cementing, casing services, mud logging, and supply of general consumables? If incoming operators are required to use the local service providers in the region, what are their availability and performance compared with acknowledged industry contractors? All these factors can influence the rig mobilization estimate.

### 6.6.2 Drilling operations

If the assets for sale are in a new theater of operations for the buyer, the drillers must scour the data room for details of drilling rig availability, including their number, type, size, specification of rig packages, capabilities, maximum water depth, BOP stack pressure rating, and reliability.

Enterprise Oil faced this challenge when operating as a relative new comer in the Black Sea in the 1990s. With no western rigs available in the region, and it being prohibitively expensive to mobilize a rig from outside and bring it through the Bosphorus Strait, it decided to use one of the stacked rigs of Petromar, the former Romanian national drilling company. Happily, the process went according to plan, and the commitment well was drilled in the Black Sea, without incident and within budget.

### 6.6.3 Drilling logistics

Napoleon famously said “an army marches on its stomach,” and rigs, wherever they may be, require constant supply to sustain drilling operations. The driller in the data room needs to ascertain what the likely challenges are for providing support to the wellsite locations by road, by sea, and by air. For example, what is the local road haulage capability? What type of air support is available or needed, fixed wing and helicopter? This is particularly important for the rapid transfer of “hot-shot” core samples to laboratories during exploration drilling, to keep rig costs down.

The driller may also be required to do a field visit to check the availability of suitable yards, warehouses, and general office support structures. While on his trip, he may check on the availability of required local service groups, such as drilling services, haulage, cranes, machine shops, and general supplies. The driller should assess the specter of local red tape regarding the general requirements and procedures relating to import and export of drilling materials and the availability of local transit agents with some idea of how long things take to be shipped in and out of the area.

### 6.6.4 Drilling environment and safety

Whatever the deal, health, safety, and environmental (HSE) concerns can never be ignored, so the driller must check the disposal requirements for fluids and low specific activity materials (e.g., scale) and review environmental considerations and protection employed on past drilling operations in the region of interest, including oil spill response capability. Following the Macondo disaster in the GOM, there is an increased regulation worldwide on well control and safety-related training given to wellsite personnel, its frequency, and its enforcement. Drillers in the PDR should include a thorough review of the seller’s inspection, testing procedures, and certification of downhole, handling, lifting, and well control equipment to gain an appreciation of the seller’s safety culture.

Table 6.3 Driller's check list for material to gather in the data room.

| Priority | Drilling and completion (D&C) data |
|---|---|
| 1 | **D&C reports (with drilled depth vs. time and cost curve) for each well in the field(s)** |
| | **D&C costs per well, with breakdown of tangible and intangible costs** |
| | **Details of future drilling campaigns** |
| | **Schematic of completed well with dimensions of open hole sections and casing strings** |
| | **Details of leak off tests, squeeze jobs, plug backs** |
| 2 | Drill bit records |
| *3* | *Daily drilling operations reports if possible* |
| | *Drilling and completion fluid properties* |

The priority data list for the driller is shown in Table 6.3.

A site visit by the M&A team driller may be deemed necessary as part of the due diligence. Such an inspection would invariably include a review of firefighting capabilities at the wellsite, the availability of breathing apparatus, and first aid equipment. A final HSE check would be a review of medi-vac capability from remote rig locations and the proximity and quality of local medical facilities and staff. If any of the above is missing or substandard, then the likely buyer will have to pay to have them brought up to and maintained at the appropriate level, which needs to be factored into the bid price.

## 6.7 Geoscientific roles

Geoscientists are defined in this textbook as a group that includes geophysicists, geologists, and geomodellers. When evaluating an investment opportunity, whether it be a prospect in some exploration acreage, a discovery under appraisal, or a producing field, their primary role is to develop an understanding of the subsurface in the region of interest, to understand the downside, and to review any missed upside potential.

All geological or geophysical interpretations are model derived, so the correctness of associated volumetric calculations is limited by data quantity and quality and the suitability of the model chosen. As mentioned in Section 1.5.1, when reviewing a seller's estimated PIIP and associated risk, one should bear in mind what is actually known and what is being assumed and whether the seller's chosen volumetric input parameters reflect these concerns. Therefore, the M&A team geoscientists must strive to understand how uncertainty in assumptions and input parameters impact on the seller's model. For example, subsurface analysis is rarely done with a perfect dataset, but all buyers require a view on whether the seller's model can answer key exploration or development questions. This often comes down the quality of data coverage, i.e., where wells and seismic surveys are located and how good are the interpreted results from them, and whether there are other alternative models that honor the data equally well or better.

In the case of an exploration venture, any farm-in will require the buyer paying for some or all the cost of drilling an exploration well in order to gain equity. Hence, a full understanding of the prospect is necessary. To this end, sufficient data should be collected from the PDR in order that a detailed PIIP estimation can be performed and allow risk and potential upside and downside to be factored into the calculations. Exploration well drilling costs will also need to be verified. Armed with a better understanding of the prospect(s) on offer, the geoscientist can focus on assessing the quality of the seller's work.

Let us now consider the contributions from the various geoscientific roles in the M&A team.

### 6.7.1 Role of the geologist

Geologists tend to focus more on the exploration and appraisal of discoveries in a seller's portfolio rather than on producing assets. Hence, they need to acquire a broader picture of the seller's acreage on offer, so understanding the geological evolution of the basin(s), tectonic history, and regional stress orientation is useful. They need to become familiar with the assets' reservoir potential, via any diagenesis and sedimentological studies (i.e., for sediment transport and input direction). Concerning the exploration acreage on offer, subsurface due diligence will include a review of the seller's proposed source potential and migration paths and timing, the proposed trapping geometries and mechanisms, thoughts on structural trends and fault sealing, quoted reservoir porosity types (primary, secondary), and stated ranges of variability in other reservoir parameters. This will permit the buyer's geologists to develop their own independent models of geological risk that can be factored into calculations (see Appendix 3).

Capen (1976) outlined a workflow that M&A geologists can follow when auditing a seller's volumetrics and risking work to gauge its accuracy. Initially, the provided PIIP model should be reviewed in detail to see if there may be alternate scenarios to those proposed by the seller. The remaining subsurface unknowns need to be collated and a judgment made on whether they can have a serious detrimental impact on the venture. Note that if the seller provides one map only, the M&A geologist must review it with the goal of defining regions of high and low certainty.

If the seller's Monte Carlo simulation model uses non-traditional distributions that are neither normal nor lognormal, then this may constitute a red flag to the buyer (see Appendix 1). The M&A geologist must review the seller's low and high PIIP estimates, which are usually taken to be synonymous with the P90 and P10 values from a continuous probability distribution, respectively, and examine the associated P99 and P1 values to verify their credibility. This confirmation of reasonableness is based on experience and from reality checks with analogs. It may transpire that the seller's P90 and P10 values of PIIP may need to be adjusted to get more realistic P99 and P1 values. Research suggests that the primary cause of overestimated volumes in a seller's exploration portfolios is that the input low-side estimates are too high. For example, the P90 value of prospect's PIIP should be roughly the smallest volume that an operator would consider completing the well and conducting a flow test. By the same token, the P99 value of prospect's PIIP should equate to a subcommercial discovery, but with enough volume to barely sustain hydrocarbon flow to surface (Rose, 2001).

A sensitivity study should be performed on the seller's model input parameter ranges and distributions to see their influence on the resultant PIIP, which will hopefully provide a result that the buyer is comfortable with. If the seller provides documentation of how the assets' PIIP estimates have been developed and changed over time, then the buyer can interrogate this history in order to track the seller's exploration and appraisal efforts to see if lessons were learnt and whether subsurface predictions have improved.

Ultimately, the M&A team geologists must assess the asset portfolio on offer and verify how realistic is that the seller's proposed work program and associated costs to realize the claimed petroleum potential of the deal.

### 6.7.2 Role of the geophysicist

Sat at the workstations in the PDR, the geophysicists will review the assets' interpreted structure maps, showing wells and seismic line 2D grids or 3D coverage. They will scrutinize how the seismic surveys were conducted, how data were acquired, and what processing parameters were employed. Previous velocity modeling and how it ties in with well logs, vertical seismic profiles (VSPs), and well check shot data will be carefully reviewed. The geophysicists will pass judgment on the structural mapping of the assets and how uncertain they think the estimates of depth are, as this will provide the range of gross volumes used in the PIIP calculations. The geophysicist can request for workstation screenshots to be printed, to capture the essential well ties, seismic line intersections, and interpreted surfaces, but the seller usually limits the number of screenshots that can be printed and taken out of the PDR.

The M&A team geophysicist must request access to all well velocity control data in the data room. If these are not forthcoming from the seller, then significant doubt must be cast on the depth accuracy, and subsequent reservoir structure mapping by the seller cannot be fully certifiable. If VSP and check shot data are known to have been recorded in some wells, but are not present in the data room, then this is a major red light for the due diligence process (see Section 8.10).

### 6.7.3 Role of the geomodeler

Often sat alongside the buyers' geophysicists in a PDR, are the geomodelers, who are tasked with checking the quality of the seller's integration of the geophysical, geological, and petrophysical observations that have been used to build "representative" 3D numerical models of the hydrocarbon accumulations. These geomodels, which are now commonplace in the PDR, are used to calculate PIIP and to provide the static input grid for the subsequent dynamic simulations of potential recovery.

#### *6.7.3.1 Reviewing the 3D geomodel workflows*

The 3D geomodeling workflows will vary depending on the stage of development of an asset. For example, in the exploration and appraisal phases, geomodeling will be basic, as there tends to be limited input data. The results of any future delineation or development wells derived from these early geomodels may have been done with an inappropriate geological concept, and so the predicted results should be treated with care. In contrast, with producing fields, the geomodel should have a geological concept driving it and, having far more input data to constrain it, the model should provide better results.

The seller's geomodel should also integrate production data to better understand reservoir connectivity. For example, when dealing with heterogenous reservoirs, the seller should employ a facies model to define flow units and barriers, but this is less important for homogeneous reservoirs. If the geomodel's 3D grid is used for PIIP calculations, it must be of high quality; honoring input surfaces, well tops and faults and have no negative cell volumes. If the grid is subsequently used for dynamic reservoir simulation, the grid cells should be regular and have little or negligible twisting, and the 3D grid should be aligned with the main flow direction to limit permeability anisotropy effects.

There are different workflows for different reservoir lithologies; carbonates often require modeling and characterization of fractures, whereas sandstones may need diagenesis modeling to spatially alter reservoir properties.

#### *6.7.3.2 Checking the 3D geomodel*

While in the PDR, geomodellers are encouraged to perform a rapid check of the static model provided on the workstation by the seller. For example, for an oil field, the geomodel should be interrogated to provide the following four outputs: net reservoir volume (NRV), net pore volume (NPorV), hydrocarbon pore volume (HPV), and PIIP. One can then back calculate the following average geocellular rock and fluid properties for oil leg of the reservoir in the geomodel, so that

1. NTG = NRV/GRV
2. porosity = NPorV/NRV
3. oil saturation = HPV/NPorV
4. oil formation volume factor = HPV/PIIP

These four average properties can then be compared with the corresponding values derived from petrophysical analysis in key wells to check for agreement (Shepherd, 2009). A key well is one that controls relatively more of the reservoir volume, has good log data that are usually calibrated to core, and so has a major influence on the quality of the modeled PIIP. If there are major divergences between the geomodel and the key wells, then this should alert the M&A team that there may be an issue with claimed PIIP.

Regarding a geomodel's volumetric uncertainty, due diligence must ascertain whether the defined uncertainties surrounding structure and fluid contact depth represent the uncertainty across the entire field. Poor well log coverage compared with the mapped hydrocarbon area should be expressed via higher uncertainty in computed volumes. As mentioned in Section 1.3.4, one should always verify that the resulting volumetric spread is reasonable, as wider P10/P90 ratios are expected for exploration and appraisal phases (Rose, 2001). The M&A team geomodellers should also perform their own 'back of envelope' volumetric estimates of the seller's static model best technical estimate, which is often assumed to be the 2P case, to reality check and compare with seller's reported volumes.

A PDR workflow is provided in Fig. 6.2 to help review complex 3D geomodels.

The following checklists correspond to the 3D geomodel review workflow.

The basic facts about the 3D geomodel should include

- Description of the seller's workflow used to build the 3D model.
- What depositional environment and geological concepts have been assumed?
- Have cross field well correlations been done?
- Have these well correlations been employed as inputs to reservoir zonation?
- What reservoirs are modeled?

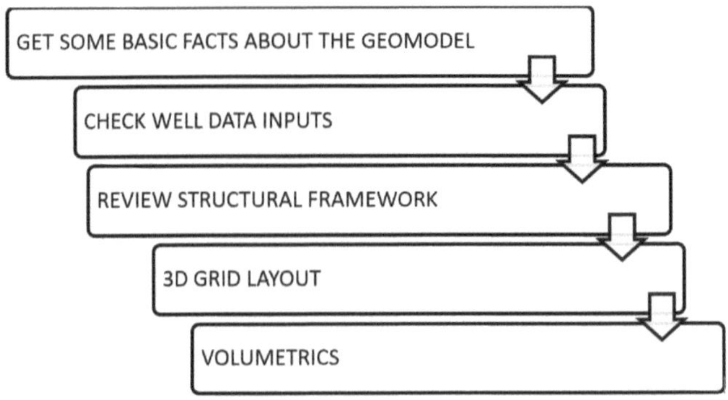


**Figure 6.2** Workflow to verify the essential elements of the seller's 3D geomodel. *(After Ålvang 2016.)*

- What types of trap are assumed in the model?
- Does the model assume single or multiple fluid contacts?
- What data are the fluid contacts based on?

Well data inputs must be verified as follows:

- Is the ratio of well coverage versus the field area reasonable or does the field appear to be under appraised?
- Have core data been used to constrain the model?
- Have pressure data been used to constrain the model?
- Are there any well logs missing from the 3D model?
- How do the petrophysical property ranges in the 3D model compare with those calculated in key individual wells?
- Have geological trends been imposed in the 3D model which have caused petrophysical property distributions to diverge from those seen in the wells?
- Do the zones and surfaces in 3D model match those seen in the wells?

The structural framework of the 3D model needs to be reviewed to confirm the following:

- Is the seismic horizon in the 3D grid comparable with the input horizon?
- Are the internal zones and flow units within the model realistic?
- Are the fault patterns modeled correctly?
- Are the faults in the model assumed to be sealing or nonsealing?
- If the reservoir is defined by structural closure, do the fluid contact ranges honor it?
- Is the structural closure of the model within the 3D grid area?

The 3D grid layout needs to be confirmed as follows:

- What is the horizontal grid increment?
- What type of layering method has been used in the 3D model?
- How does the thickness of the vertical layering in the geomodel compare with that observed in cores and logs?
- Is the assumed geological concept driving the 3D model?
- If geological trends are used in 3D model, what are they based on?
- Is "twisting" of the 3D grids evident?

Finally, the M&A team geomodeler should fully understand the seller's volumetric cases:

- How is the volumetric base case defined?
- How does the seller's original PIIP case compare with the eventual best technical case, and what has been altered during the derivation of the latter?
- What is the seller's stated range of volumetric uncertainty?
- What are the major sources of uncertainty in the seller's volumes?

### 6.7.4 Common errors of in-place volumes

Industry studies have repeatedly shown that the largest cause of overestimation of PIIP volumes is badly characterized GRV. This arises often from the reservoir being poorly defined both areally and vertically, via lack of well penetrations and limited seismic coverage, and inaccurate seismic mapping and depth conversion. Where exploration is concerned, the prospect structure may have had insufficient hydrocarbon charge or maybe its seal is imperfect so that the structure will not be full to the spill point. Errors may be due also to poor geoscientific practice which leads to incorrect reservoir areas and volumes being calculated. For example, when dealing with structural traps, the geophysicist will map a horizon below which one can compute rock volumes. However, it is often the case that the mapped horizon is not the top reservoir surface. These errors are subsequently compounded by using unconstrained and arbitrary input parameters to estimate PIIP. These issues are discussed in Appendix 1.

Due to the relative importance of the GRV parameter in the PIIP calculation, while in the data room, the geoscientists must quickly check not only the seller's PIIP values but also the method used to calculate them. While most GRV calculations today are performed with sound algorithms and using area-depth plots for the top and base reservoir surfaces, one can use the simple "slab formula" of "reservoir thickness x area" to obtain a rough and ready estimate of trap volume. This approach is quick, but it is approximate and consistently overestimates GRV, particularly where reservoir thickness is greater than vertical closure. The corollary is that the errors introduced via use of the slab method are minimized when used to model thin reservoirs over large areas with significant vertical structural closure. If the seller provides a GRV estimate from the slab formula, buyers can attempt to correct it by using a geometric shape factor.

The correction for shape is calculated from the structural relief of the trap, which is called the height of closure, and is the vertical distance between the crest and the spill point. The height of closure is then input to a relationship chart (Fig. 6.3). The more appropriate gross sectional shape of the reservoir is selected, such as either a rounded dome (half a ball) or a flat-topped container (soup dish), and the corresponding correction is obtained.

If time permits, and maps are provided by the seller, the M&A geoscientist should construct area versus depth plots for the accumulation under investigation. These plots are straightforward to perform on computer, for which several low cost and easy software applications exist, and the more sophisticated approach gives greater accuracy in the GRV estimate. An example of an area—depth plot is shown in Fig.6.4. Having such a plot to hand allows the M&A team to perform a rudimentary check on the seller's planned "optimum" well locations and any forecasted up-dip potential.

The priority data list for the geoscientist is shown in Table 6.4.

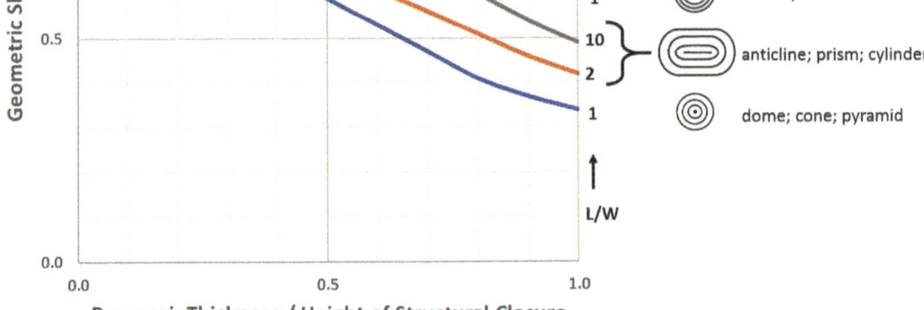


**Figure 6.3** Geometric shape factor chart, with corrections for different reservoir shapes derived from height of closure. *(After Logicom E&P Limited 2018.)*

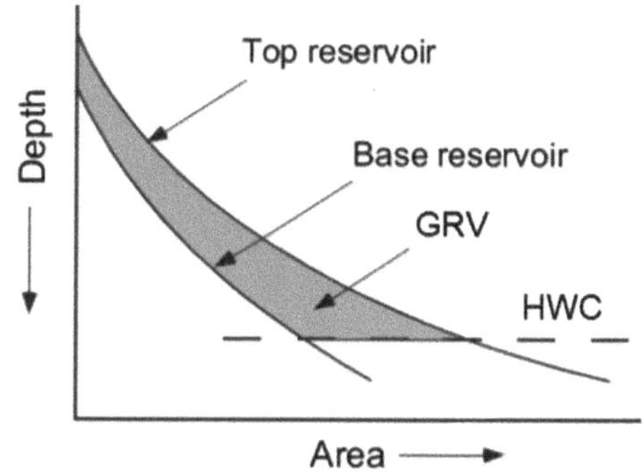


**Figure 6.4** Area–depth plot to compute gross rock volume. *(After Demirmen 2007.)*

**Table 6.4** Geoscientists' checklist for material to gather in the data room.

| Priority | Geological, geophysical, and geochemical data |
|---|---|
| 1 | **Geomodels, Monte Carlo models, PIIP estimates** |
| | **Basin modeling studies, prospectivity studies, risking methodology** |
| | **Structure maps, isopach maps, parameter maps (porosity, permeability, gross thickness, net thickness, NTG, water saturation)** |
| | **Structural modeling and stratigraphic cross sections** |
| | **Seismic interpretations (time, depth), velocity models** |
| | **Composite logs, End of Well Reports** |
| 2 | Imaging and dipmeter data |
| | Previous depositional/environmental studies |
| | Whole core descriptions |
| | Past studies of all types (geochemistry, biostratigraphy, petrology, etc.) |
| *3* | *Detailed petrology (SEM, thin sections, formation water analysis)* |

## 6.8 Role of the petrophysicist

Log analysts help the M&A team to ascertain PIIP as accurately as possible and provide support on recovery factor (RF) estimation. Where exploration opportunities are concerned, they assess the reservoir potential of any permeable formations present in the region. They also provide essential validation of the petrophysical parameter ranges and distributions that are input to the geomodel.

An immediate priority of the petrophysicist in an M&A team is to access digital data for all wells: logs, core, wireline formation pressures, and deviation surveys for depth conversion. Details of the nature of fluids sampled in the wells, and from what depth and formation, can help verify the seller's assignment of fluid saturation in the geomodel and check assumed hydrocarbon water contacts.

Gathering data is less straightforward if the material is not in the English language or are "vintage" (i.e. non-combinable tools with unfocused measurements requiring more logging runs in the wells, such as those run in the former Soviet Union). These factors require language translation, tool response modeling, and increased depth matching; adding time, cost, and potential error. Issues of log data of different vintages and from different logging contractors need to be resolved, which is possible if a "marker formation" is present above the reservoir of interest. This marker is a thick bed of known lithology in which logging tool responses can be checked. In addition, there may be a requirement to reconcile possible depth discrepancies between the more accurate wireline logs and those run on drill pipe, as failure to do so will cast doubt on subsequent geomodeling.

A major issue is to develop an understanding of the petrophysical methods and associated parameters that have been used to produce the asset's "holy trinity" of porosity, water saturation, and net-to-gross (NTG), which directly influence calculated PIIP. To this end, the petrophysicists must investigate shale indicators, porosity tools, and computation methods, compare core data with logs, and ascertain whether the correct overburden correction has been applied to core data. The latter ensures that surface measurements of core properties are corrected to reservoir conditions. They must ensure the appropriate water saturation calculation methods have been applied, check sources of formation water resistivity, and what cut-offs were used to compute net sand, net reservoir, and net pay.

Now let us consider the elements of PIIP on a parameter by parameter basis.

### 6.8.1 Shale volume

It may appear odd with the advent of unconventional resource developments targeting shale plays, but traditionally, shale is regarded as nonreservoir in conventional reservoirs. The estimated volume of shale is used to correct porosity and water saturation for the effects of clay bound water, which reduces pore space available and makes reservoirs appear wetter than they actually are. The PDR petrophysicist needs to confirm which shale indicators have been used, whether the estimated shale volume has been checked, and by what means. Standard practice is to make a number of shale volume estimates using different logging tool responses, and then selecting the lowest estimate. Shale volumes can be cross-checked against measurements in conventional core or sidewall core samples. A less sophisticated way is to compare with wellsite descriptions of drill cuttings; caution is advised, as one never knows exactly from what depth in the borehole the cuttings came from. As shales are often washed out during the drilling process, a simple check is to compare the caliper log with the drill bit size to highlight any borehole enlargement and infer shale zones that way.

### 6.8.2 Porosity

After using shale volume to differentiate between what is reservoir and what is not, petrophysicists will turn their attention to porosity, which is the percentage of pore volume that can contain fluids. One immediate question to ask if whether the seller's analysis has used effective porosity or total porosity. The former, which has been corrected for shale volume, relates to interconnected pore volume contributing to fluid flow and excludes isolated pores. The latter relates to total void space in the reservoir, irrespective of whether or not it contributes to fluid flow. Remember that effective porosity is always lower than total porosity.

It is useful to confirm which logging tools have been used to "derive porosity"; no logging tool can measure porosity directly, so it requires a model to back out porosity from its response. The density tool is usually preferred, being a pad device which is pushed against the wall of the borehole, so reducing the effects of drilling mud on the tool response. However, the density tool is badly affected if the borehole is rugose and washed out; the tool also over predicts porosity in gas-bearing zones. The neutron and the sonic tools are the other mainstays for deriving porosity; these are mandrel type tools that stand off from the borehole wall and are therefore affected by the borehole environment, and usually need correction before analysis. Whichever tool is used, any porosity estimate should be compared with core data which are considered to be the "ground truth." One should confirm whether the core plug measurements have been corrected for compaction effects and to check which correction method was used.

Besides the methodology used, the parameters assumed in the log analysis must be reviewed. For example, to estimate porosity from the density log, while ignoring log response corrections for shale volume and gas effect, the total porosity from the density log is equal to (DL-DM) divided by (DF-DM), where DL is the density log reading, DM is the reservoir matrix density, and DF is the reservoir fluid density (usually assumed to be 100% water for quick-look analysis), all in units of gm/cc.

The matrix density should concur with the grain density measured on core, which is also a check on whether the correct lithology has been used in the porosity calculation. The fluid density from log-to-core calibration should be physically realistic, i.e., the assumed fluid density should lie between mud filtrate density and reservoir fluid density.

### 6.8.3 Water saturation

Water saturation is considered to be the most challenging of the petrophysical parameters to compute as it can be derived in several ways which introduces uncertainty and can lead to disagreement. Water saturation is conventionally calculated using the equation of Archie (1942), which is an empirical quantitative relationship between porosity, electrical conductivity, and brine saturation of rocks, or by using a shaly sand variant of the equation.

The buyer's petrophysicist must confirm which method the seller has used; clean sand (Archie), shaly sand, or multimineral. The shalier the reservoir rock is, and the fresher the formation water is, the less suitable the Archie equation becomes, until more complex methods are needed (Fig. 6.5). Higher shale volume and lower formation water salinity implies more uncertainty in the computed water saturation.

The provenance of the parameters used with the Archie equation, namely the Archie coefficient (a), the cementation exponent (m), and the

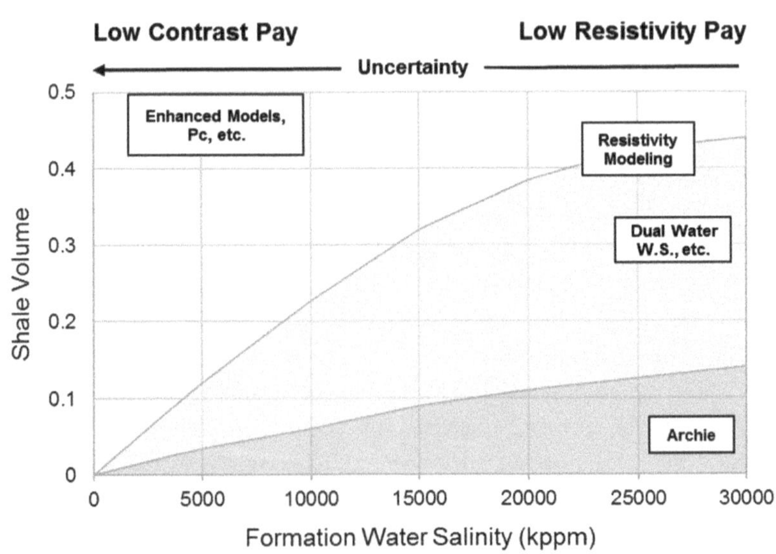


**Figure 6.5** Appropriateness of water saturation calculation methods versus shale volume and formation water salinity. *(After Heavysege 2002.)*

saturation exponent (n), should be confirmed. For example, if the rock is oil-wet (as many carbonates are), then the saturation exponent should be higher than its normal value of around 2.0. Water saturation measurements using the methodology proposed by Dean and Stark (1920) on core samples from wells drilled with oil-based mud provide a good check on log-to-core data consistency.

Along with an appropriate, environmentally corrected deep resistivity log reading, another crucial input parameter to water saturation calculation is the formation water resistivity (Rw). The PDR petrophysicist must verify that the Rw values are consistent from different sources such as logs, produced water from flow tests or formation sampling, or inferred from a formation tester water gradient. If Rw is not consistent, check whether the Archie cementation coefficient is correct, and whether salts other than sodium chloride (NaCl) are present.

Auditors should ensure that they record perforated intervals, and produced rates of oil, water, and gas so that they can verify their interpretation of what fluids are located where in the wells and so cross-check hydrocarbon water contacts (HWC) in the field. If there are disagreements, this might suggest the casing cement job is suspect, and is not providing hydraulic isolation of producible formations.

### 6.8.4 Net-to-gross

After checking shale volume, porosity, and water saturation, the last piece of the petrophysical puzzle for conventional reservoirs is the computation of NTG.

The PDR petrophysicist should always check what the seller is referring to when the term "net" is used and then confirm what cut-offs have been used to compute net pay intervals in the wells. Three different cut-offs tend to be applied in a cascade manner (Fig. 6.6) to a conventional sandstone reservoir to compute various net values, as described below:

1. A lithology cut-off, based on shale volume, is applied in the first instance to remove the shale prone layers so that only sand remains. This provides a net sand value.
2. An effective porosity cut-off, usually taken from a permeability–porosity relationship, is then applied to remove the poorer quality, tighter rock so that only the more productive reservoir remains. Porosity is used as a proxy for permeability as the former tends to be available in every well. The cut-off permeability is often assumed to be 1 mD for oil reservoirs and 0.1 mD for gas reservoirs and these values are input to a permeability-porosity relationship to determine corresponding cut-off

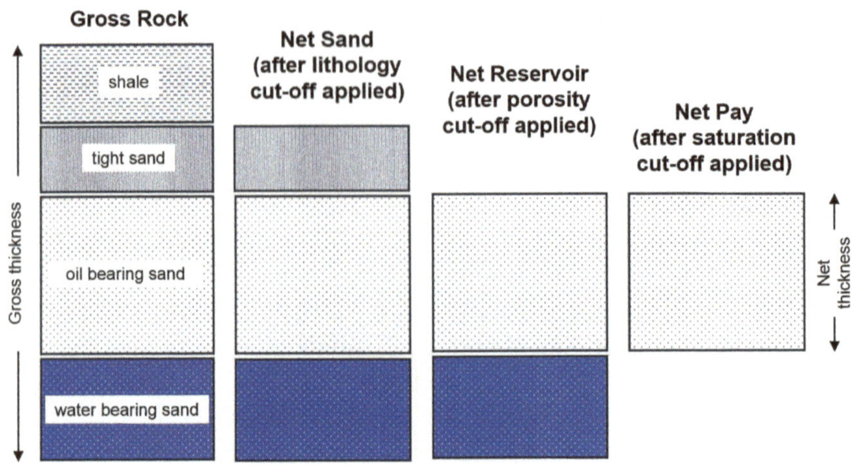


**Figure 6.6** Impact of applying a cascade of cut-offs to calculate net pay of a conventional oil-bearing sandstone.

porosity values. These arbitrary values are taken to approximate to the lowest permeability that could yield hydrocarbons in favourable conditions within a field's lifetime. This provides a net reservoir value.

3. Finally, an effective water saturation cut-off, usually based on hydrocarbon—water relative permeability (Kr) curves, is applied to remove the wetter net reservoir rock that may not flow hydrocarbon, so that only hydrocarbon-bearing net reservoir rock remains, which is termed "pay." The saturation cut-off should be cross-checked with the relative permeability curves in the dynamic simulation model. This provides a net pay value.

However, when dealing with fractured carbonates, the application of lithology and porosity cut-offs may be meaningless, and an arbitrary saturation cut-off may be the only way to derive an NTG value. The cut-off criteria may have to be revisited if reservoir zones regarded as net pay by the seller have experienced a number of failed formation tests where no fluid could be produced.

### 6.8.5 Fluid contacts

Once the petrophysical interpretations have been checked for the various key wells in the structure, the physical location of the hydrocarbons within the accumulation needs to be verified as part of the PIIP due diligence. This entails confirming the depth of the HWC from logs, hopefully supported by formation tester pressure gradient data. Then the function relating water saturation to height above the free water level (FWL) in the reservoir must be reviewed; water saturation should be 100% at the FWL, decreasing with increasing height above the fluid contact, which defines the transition zone from which both hydrocarbon and water can flow, and ultimately reaches the connate region with the lowest water saturation, where only hydrocarbon will flow. Any assumed contact must be cross-checked with observed well test results to ensure consistency.

The formation tester pressure data are crucial in determining the location of the FWL, which is usually taken as the intersection of the hydrocarbon gradient and the water gradient. Analysts often rank the data provided by the formation test mobility values, prior to drawing their linear gradients, and may only consider formation tests with mobility values greater than 50 mD/cp say, to be used for gradient derivation. Mobility is the ratio of effective permeability to phase viscosity, which in many cases will be the mud filtrate viscosity. One should always ensure that the true vertical depth (TVD) subsea is recorded and used in the plot of formation test pressures. A simple quality control technique is to plot the measured drilling mud pressure gradient alongside the formation test data to check for supercharging in tight formations, where the measured pressure is boosted to an erroneously high value. Only linear gradients should be drawn through formation test pressure data.

### 6.8.6 Saturation height function

Having confirmed the FWL or boundary between the hydrocarbon and the underlying water legs, the petrophysicist can check the derived function of water saturation versus height above the FWL, by comparing it to the water saturations interpreted in each of the key wells on the structure. Even if the seller claims there is a good relationship in the IM, PDR petrophysicists must verify such sweeping statements for themselves.

The appropriateness of the saturation height function (SHF) in the geomodel must be checked for the assumed hydrocarbon and rock type, or merely to verify if one was used at all. All proposed fluid contacts must be substantiated, which is no mean feat when faced with structure maps showing only oil-down-to (ODT) and water-up-to (WUT) depths, and the contact might be anywhere in between. Some companies assume the fluid contact is half-way between ODT and WUT, which is a completely arbitrary approach (Harrison, 2014b).

The priority data list for the M&A petrophysicist is shown in Table 6.5.

**Table 6.5** Petrophysicist's check list for material to gather in the data room.

| Priority | Petrophysical data |
|---|---|
| 1 | **Digital well logs (including cased hole and mud logs) should include resistivity, sonic, density, neutron, gamma ray, caliper.**<br>**Routine core analysis (RCA), including Kh, Kv, porosity, grain density, Sw**<br>**Well deviation surveys for log depth to TVDSS conversion.**<br>**Field petrophysical studies, with previous well log analyses with petrophysical summaries (CPIs if available).**<br>**Wireline formation pressure summaries; formation water sample analyses.**<br>**Special core analysis (SCAL) I (Kr curves, Pc-height).** |
| 2 | SCAL II (resistivity index, formation factor, hydrostatic compaction, CEC).<br>Log header info (may not be on file) including surface locations. |

## 6.9 Role of the reservoir engineer

The reservoir engineer's primary function in the data room is to verify the seller's resources, profiles, well scheduling, and facilities requirements. In pursuit of this goal, surface production and fluid data must be related to reservoir pressures and rock characteristics to understand the drive mechanisms in the reservoir, which in turn will influence their estimate of hydrocarbon recovery from the assets and influence any planned well locations. The most common reservoir drive mechanisms and their associated ranges of RF are described briefly in Appendix 1.

Very often, the oil assets on offer will be on water flood, as it is one of the most efficient recovery mechanisms, where reservoir voidage due to fluids being produced is replaced by injecting water to maintain reservoir pressure at or near its initial level.

### 6.9.1 Water drive sufficiency

The M&A team reservoir engineer must check that there is evidence of sufficient pressure support by computing the voidage replacement ratio (VRR) and, if data are available, by reviewing the field material balance. VRR is the ratio of injected reservoir volumes and produced reservoir volumes, and it should be maintained at greater than unity. Based on the value of VRR, one can estimate whether there is enough voidage replacement or if it needs to be supplemented with more injection wells. Material balance, which uses conservation of mass to define a reservoir's drive mechanism, can show evidence of pressure support from the producing history match. The data required for material balance include the average reservoir pressure history, with corresponding dates of each pressure survey, reference depth of the pressure measurement, a note of the well in which the readings were taken, plus the production from the well to that date. As the pressures may be recorded over many years, and be measured by different service companies, with different vintages of technology, one should note the gauge type and its accuracy too. Lack of observed pressure support in parts of a field may be due to reservoir compartmentalization.

### 6.9.2 Compartmentalization

One can review possible compartmentalization by checking with the geophysicist if there is evidence of any major faulting shown on the seismic sections. Also, one can review the pressure versus TVD analysis from the formation tester on a well-by-well basis to see if there are a series of different gradients that may infer compartmentalization. It is instructive to review the formation tester surveys, run on wireline or on drill pipe, that have been run in different wells at different times in the field's producing life, as offsets in their pressure gradients would imply vertical layering due to differential depletion in the reservoir. Where the pressure versus TVD plots are difficult to interpret, such as when the fluid pressure gradients are similar, one should use the excess pressure method (Brown, 2003). The technique, which removes the effect of a selected fluid density by subtracting its corresponding hydrostatic gradient from the measured formation pressures, improves the visualization of small fluid density differences and the existence of pressure barriers.

If well test data are available for wells in the different "compartments," one can perform quick build-up pressure analyses in each of the wells to see if the extrapolated pressures coincide or not. Transient pressure analyses, especially the late time pressure derivative responses, might give some inkling as to which geological fault model can be invoked. PDR reservoir engineers should request time-pressure-rate (TPR) data (and the associated interpretation files if possible) for key wells from the seller's staff. From a regional perspective, if the well density is higher and the estimated ultimate recovery (EUR) per well is lower than seen in fields with less reported compartments, then this fact may indicate compartmentalization in the field under scrutiny. Use of these observations will help the M&A team develop a better understanding of vertical and areal sweep efficiency in the assets under investigation.

### 6.9.3 Reservoir fluids

To develop an understanding of the reservoir fluids of the seller's assets, the buyer's reservoir engineer needs to have access to laboratory reports on the fluid pressure—volume—temperature (PVT) behavior, as measured in the laboratory on fluid samples. One should note where the PVT samples came from, as they can be obtained from a variety of sources, such as surface separator recombined tests, or formation tester chamber, or bottom hole sampler, and each source has its own challenges.

If PVT reports are unavailable, the reservoir engineer should strive to get values for oil API gravity, gas oil ratio, bubble point pressure, gas-specific gravity, and condensate gas ratio in order to use published black oil correlations to carry out any reservoir engineering checks.

Gas condensate reservoirs will also need values for dew point pressure, which, when combined with the condensate gas ratio and the size of the accumulation, can be used to estimate whether dry gas cycling is a viable option for field development.

Reservoir fluid viscosity data are useful inputs for displacement efficiency studies. These can be combined with relative permeability data to give an end point mobility ratio (M), which is the mobility of water at residual oil saturation divided by mobility of oil at irreducible water saturation. This parameter allows the PDR reservoir engineer to get a feel about the potential effectiveness of water drive in the area. The lower the value of M, the better the displacement by oil by water flood, and permits a reality check of the seller's claims via comparison with the performance of nearby analog fields.

Data compiled from the data room can be checked against published PVT correlations to highlight any dubious results. The reservoir pressure and temperature values provided can be converted to hydrostatic and geothermal gradients, and cross-checked with data from drilling mud weight reports and log headers. In the absence of any field data, regional pressure and temperature gradients may be the only source of information.

### 6.9.4 Reservoir heterogeneity

To investigate reservoir heterogeneity, it is crucial to get hold of core permeability data. It is also beneficial if a production log tool (PLT) spinner survey is available from the cored well, as it will allow an independent check of the "static" permeability profile from core data versus the "dynamic" flow contribution profile from the PLT. Reservoir production flow profiles may also be obtained (albeit less commonly) from the emerging fiber optic technologies for distributed sensing in wells using temperature (DTS), acoustic (DAS), and chemical (DCS) inputs.

One should verify that the wells perform on production as one would expect from their core-based permeability profiles. If water production is observed in some of the wells, one should check the time of water breakthrough in the wells, and record which zones appear to have watered out, and verify that they concur with that expected from the petrophysical analyses and modeled production forecast.

There are a couple of useful coefficients of heterogeneity that are simple to calculate, and which can then be compared against those reported in nearby analogue fields to check asset performance. Dykstra and Parsons (1950) introduced the concept of the permeability variation (V) to describe the degree of heterogeneity within a reservoir. Values for the Dykstra–Parsons' coefficient range from zero for a completely homogeneous system to one for a completely heterogeneous system. Schmalz and Rahme (1950) also proposed a single term to characterize the permeability distribution in a reservoir, defining the Lorenz coefficient of heterogeneity as the ratio of a plotted area from the actual core data and that of a theoretical homogeneous system. The Lorenz coefficient also ranges from zero to one, where a uniform permeability reservoir has a Lorenz coefficient of zero (Craig, 1971). A more qualitative method of estimating heterogeneity is to compare the "permeability thickness" product from analysis of a well test with the arithmetic, geometric, and harmonic averages of core permeability taken in the tested well.

### 6.9.5 Well and field performance

The reservoir engineer's holy grail is getting access to individual well production histories to allow regression line fitting to forecast future well production. This is known as decline curve analysis (DCA), which provides relationships of rate against time and cumulative production and gives an EUR for the well. The build-up and production and behavior on plateau and in decline can be modeled, and forecast production profiles that correspond to a future well program can be evaluated.

Buyers' staff should beware accepting subsets of wells chosen by the seller as they may have been "cherry picked" and may not be representative of the spread of performance of wells in the asset for sale.

Better results are obtained by performing DCA at well level, rather than at field level which tends to obscure how the asset is performing and usually overestimates reserves (Harrell et al., 2004). One can place more faith in DCA results if there is sufficient history of say 2 years or more on which to perform regression. The M&A analyst should match to cumulative production rather than match to the rate, which is often reported on a monthly basis. Rates can be either calendar (produced volume divided by days in the month) or actual (produced volume divided by time onstream in that month). The former means the operational uptime is incorporated into the DCA. However, if one uses the latter preferred method, then in addition to a decline factor, an uptime factor can also be obtained from matching the cumulative produced volume.

If actual production data are not made available by the seller, or one wishes to check the seller's profiles for suggested for exploration prospects, some background research may yield analog field production data to allow reality checks on initial well rates, build-up of field production rate, annual off-take on plateau, recovery on plateau, field, and well decline rates, well density in field, recovery per well, and producer/injector ratio.

One can plot the field's average value of "mobility thickness" (i.e. the net permeability thickness divided by the hydrocarbon viscosity, $kh/\mu$), which combines reservoir rock and fluid properties, versus the claimed recoverable volumes or RF estimates. A reality check is obtained by adding published values from nearby analog fields and observing whether they coincide with the seller's claimed performance of the assets on offer.

It often falls to the PDR reservoir engineer to review the seller's numerical simulation models, so let us now turn our attention to what is required in the data room.

### 6.9.6 Initialization of the dynamic simulation model

Prior to any dynamic reservoir simulation being done, the numerical model must be primed with the measured static properties. This section outlines the workflow in checking the model's initialization, shown schematically in Fig.6.7.

It is important to confirm that the grid covers a sufficient area and that if an aquifer is present, it is included in grid. Note that the grid for a full field model (FFM) will be coarser than that required for a sector model. To capture heterogeneity adequately, both spatially and vertically, the grid block dimensions must be such that there are a reasonable number of cells between the wells, and there must be sufficient layers, with the appropriate upscaling having been applied. Faults are not always vertical, and they must be properly represented by the grid.

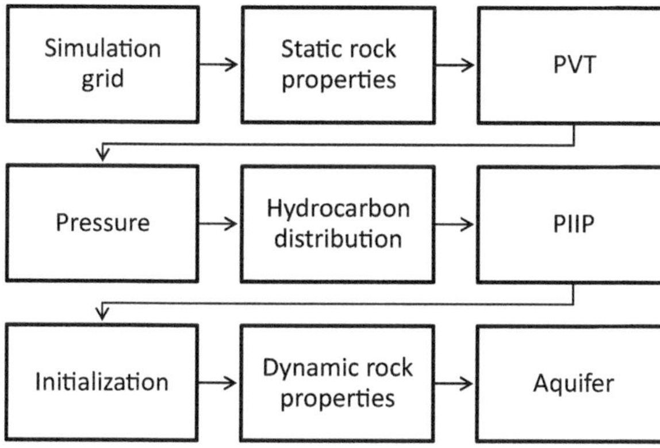


**Figure 6.7** Workflow to check the initialization of seller's dynamic simulation model. *(After Quinn 2016.)*

One should confirm the provenance and the validity of the static property inputs, including porosity, NTG, horizontal permeability, and vertical/horizontal permeability ratio. The permeability upscaling is a crucial step, and a comparison with permeability measured on core and derived from well test analysis is a necessary reality check. It is also important to confirm that the rock compressibility assumed in the model is reasonable for given lithology, depth, and porosity.

The M&A team reservoir engineer must confirm whether the PVT model used is black oil or compositional, and whether it is appropriate for the reservoir being modeled. The PVT reports provided should be reviewed to check the validity of the fluid samples and confirm that the simulation model input is correct. If areal and vertical variations of PVT properties exist, they should be modeled. One should assess how the black oil PVT tables are generated, and whether they are matched to the appropriate facilities conditions in the field. In the case of gas condensate, one should satisfy oneself that the compositional model has been adequately matched using an equation of state. Finally, it is important to ensure that the formation water properties of viscosity and compressibility have been entered correctly.

Regarding the initial pressure in the model, one should verify that the recorded pressure data are converted to the correct datum. A useful check is to confirm that the reservoir pressure gradient in the model matches that measured by formation testers in the wells.

A crucial step is to check the hydrocarbon distribution in the static model, especially to confirm that the various fluid contacts are defined correctly and honor any observed areal variations. As in the geomodel, the

SHF should be reality checked against water saturations derived from logs. The SHF is directly related to the capillary pressure input in the model, which, in turn, is related to the relative permeability curves. The dynamic property data need to be checked for shape and end points, and whether they are appropriate for the assumed reservoir wettability. If an aquifer model is present, it should be checked to ensure its parameters are sensible.

Finally, the PIIP should be cross-checked against that from the geomodel and there needs to be a review of any adjustments that were made to make them agree. On initialization, the simulation model should be stable, i.e., at equilibrium.

### 6.9.7 History match and prediction from the dynamic simulation model

Once the model is initialized correctly, the simulation runs can commence to match the observed dynamic behavior and production history. This section outlines the workflow in checking the model's dynamic elements, shown schematically in Fig. 6.8.

The following checklist corresponds to the history match workflow (Fig. 6.8) which begins with the seller's dynamic model history match parameters being verified as follows:

- If the seller has changed the model pore volume to achieve a history match, the buyer should determine whether the changes are regional, which is acceptable, rather that local "fixes," which is more of a concern.
- If specific modifications have been made to the wells in the model, the buyer should ask himself if the changes are geologically reasonable and within acceptable engineering limits.

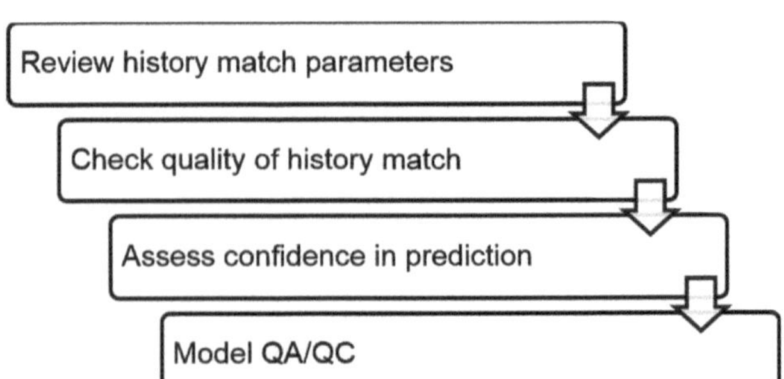


**Figure 6.8** Workflow to check the history match of seller's dynamic simulation model. *(After Quinn 2016.)*

- The parameters that are often altered include horizontal and vertical permeabilities, relative permeability curve shape and end point values, reservoir/well, fault, and reservoir/aquifer connections, and the aquifer size and strength.

The PDR reservoir engineer must then assess whether the history match is of acceptable quality and constrained by a good historical dataset as follows:

- History matching is traditionally controlled by the primary hydrocarbon phase production rate, usually oil or gas, but how does the match hold up when considering other phase rates and ratios, such as GOR, water cut, CGR, and WGR?
- Does the history match improve or worsen if one zooms in from field level down to well level?
- Have the injection rates been honored at the well level, or are they input to the model at field level?
- The history match must honor observed pressures, so do the model results compare favorably with downhole reservoir pressures recorded initially by WFTs, then by regular static pressure measurements, and during subsequent PLT surveys?
- In order to confirm that the vertical lift and inflow performance relationships for wells are adequate, one should review the model's corresponding wellhead pressures match.
- If specific well lift curves are used, rather than generic curves, have they been calibrated properly to test data, and do they cover the entire expected operating range?

Once the dynamic model is history matched adequately, it can be run in prediction mode to forecast future performance and recovery. The PDR reservoir engineer must check the dynamic model's predictions as follows:

- Are the well schedule and facilities modifications within the model are consistent with the agreed field development plan (FDP)?
- Are well groups and pipeline networks correctly modeled?
- Are the field operating conditions, such as pressure and rate constraints, for both production and injection, as well as well and plant uptimes correctly specified and applied?
- Are the new well locations in the dynamic model consistent with the seller's FDP?
- Do the new wells' connection factors reflect their planned trajectory and, in the case of high rate horizontal wells, does model recognize that wellbore friction may be an issue and is it modeled correctly?

- During the prediction runs, one should confirm that the mode of well control within the model is appropriate and applied. For example, are well cut-offs applied such as minimum oil rate and maximum water cut, and are they appropriate?
- Are future well workovers input manually to the model or are they automatically assumed by the software, and are they practical?
- The duration of the prediction run should be checked for appropriateness, and the regional RFs should be assessed for reasonableness.
- The remaining hydrocarbon distribution needs review to determine whether there are remaining pools that might be targeted.
- With particular regard to compositional models, primarily used in gas condensate simulation, the predicted mass recovery should be checked in addition to oil and gas recovery.

Finally, the PDR reservoir engineer must audit the overall quality of the dynamic simulation forecast which may be done via a review of the modeling report only, or an inspection of the simulation files (e.g., input, simulation report, recurrent data, and map data).

The priority data list for the reservoir engineer is shown in Table 6.6.

## 6.10 Role of the production engineer

The production engineer, sometimes called the petroleum engineer, in the PDR will gather information on the existing field production operations; especially any evidence to confirm the seller's claims about well productivity and injectivity. The usefulness of such knowledge is not restricted to producing assets; understanding the main operational challenges in the region will help in assessing potential discoveries and pending development projects.

### 6.10.1 Well productivity

When reviewing production operations, it is always useful to ask the seller how many development wells turn out to be dry holes or poor producers. This line of questioning can be extended to inquire about the causes of the poor well productivity. For example, is formation damage, which is commonly referred to as "skin," the cause of the well performance impairment? Often, skin is due to pore plugging or clay swelling in the reservoir from the use of incompatible drilling and completion fluids.

If the assets being evaluated produce from unconsolidated and friable formations, there is likely to be a need to constrain well flowrates to control

Table 6.6 Reservoir engineer's check list for material to gather in the data room.

| Priority | Oil, gas, water production |
|---|---|
| 1 | **Monthly well production of oil, gas, and water volumes and injection records.** |
| ***Fluid properties*** | |
| 1 | **Original lab reports on reservoir fluid PVT, composition for all test types, i.e., constant composition expansion constant volume depletion (CVD), retrograde condensation, slim tube.**<br>**As above for injection fluids (i.e., gas, water, polymer).**<br>**Laboratory separator test results (original report and interpretations).**<br>**Water analysis (composition, density, viscosity, compressibility).** |
| 2 | Summaries of previous PVT or fluid property treatments.<br>Oil chemistry/physical properties analysis. |
| ***Core data*** | |
| 1 | **Routine core analysis (horizontal and vertical plug permeabilities, porosity, grain density).**<br>**Special core analysis (relative permeability curves, capillary pressure–water saturation, compaction effects on porosity and permeability).** |
| ***Pressure data*** | |
| 1 | **Production tests (flowing bottom hole pressure and temperature (FBHP, FBHT), corresponding flowing wellhead pressure and temperature (FWHP, FWHT), choke size, surface flow rates).**<br>**Wireline formation tests, drill pipe conveyed formation tests,**<br>**Drill stem tests.**<br>**Transient well test data (XRR files for build-up, drawdown, pulse).**<br>**Static reservoir pressures (with date, shut-in time, gauge type, depth of measurement).** |
| 2 | Results of transient pressure analysis, shut-in pressure surveys, pressure gradient surveys. |
| ***Reservoir studies*** | |
| 1 | **Check the basis of all input assumptions to the simulation model.**<br>**QC of history match and changes required for simulator match.**<br>**Check the basis of all input assumptions to the material balance model.**<br>**QC of material balance match.** |

sand production, which can erode tubulars and surface equipment. It should be noted that fields with seemingly competent formations can produce sand in later life, such as when there is water breakthrough at the oil producers, which can cause destabilization of the reservoir sand. Also, in mature gas fields, the depletion of reservoir pressure means that the overburden is increasingly supported by the reservoir structure itself, which can lead to compaction and subsequent production of fines. This late life sand production impairs gas well deliverability by filling up the wellbore with sand and causing erosional issues if produced to surface.

As sand control is often improved with appropriate completion and perforating strategies, the M&A production engineer should ask the seller what sand monitoring and control methods have been tried and put in place. A prudent operator will have recorded or monitored sand production throughout the field. There are several techniques available, from the downhole methods of noise logging using wireline and sand fill tags by slickline to inserting sonic probes and erosion coupons in surface flowlines and recording the volumes of bottom sand and water (BS&W) cleaned out of the separator during the regular flow tests for allocation of well production.

With regard to the seller's well completion philosophy, if wells have been gravel-packed or had downhole sand screens or slotted liners installed in them, then their productivity will be impaired. Some operators inject resins into the friable producing formation to stabilize the sand and prevent it being produced, but this also impairs well deliverability.

Concerning the seller's perforating strategy, the PDR production engineer must determine what was used and is it optimal for the asset in question. For example, whether the guns were conveyed downhole via wireline or tubing will play a part in the well deliverability. The former method of gun conveyance is quicker and cheaper but delivers smaller charges downhole. The latter delivers bigger charges to give larger diameter entrance holes in the casing and deeper penetration to hopefully get beyond any zones of formation damage but requires a deeper (and costlier) sump to be drilled below the perforated interval in which to drop off the spent guns. Perforating at high underbalance is now considered less important, but a constant, managed pressure drawdown should be maintained to try and remove formation damage. As important is whether the seller has selected the most effective shot phasing and density to achieve optimal well deliverability.

If the seller has attempted to stimulate the poorer performing wells in his field, by adding perforations, acidizing, or hydraulic fracturing, what was the impact on their productivity (or injectivity)? Poorly applied stimulation can hinder as well as help, so it is worthwhile checking that the applied stimulation methods did not cause sand control problems.

The PDR production engineer needs to assess whether the poorly performing wells need to be worked over, or to be redesignated, or in the worst case, to be plugged and abandoned (P&A), so the seller's past operations experience in the field needs critical review. It may be possible that a poor oil producer that is located on the flank of a reservoir structure can be converted into a water injector, to provide pressure support and improve sweep efficiency.

Thinking of possible upsides in mature developments, buyers need to assess whether any of the abandoned wells can be reentered, worked over, and brought back onto commercial production?

### 6.10.2 Oil production operations

Oil well productivity impairment can also be due to the nature of the produced fluids. For example, the seller should be asked whether there has been inorganic scale formation due to production of incompatible formation and injection waters. Calcium carbonate scale ($CaCO_3$) can be removed with an acid wash, whereas barium sulfate ($BaSO_4$) and strontium sulfate ($SrSO_4$) scales cannot and will need reperforating or sand jetting to recover the lost productivity, followed by scale inhibitor dosing to stop it from returning.

If the asset under investigation contains heavy oil, which is often defined as liquid petroleum of less than 20°API gravity or more than 200 cP viscosity at reservoir conditions with a waxy, asphaltic, dense, viscous nature (similar to treacle), then buyers should beware. Heavy crudes tend to deposit wax and asphaltenes in the production tubing, which inevitably impairs oil well deliverability, resulting in higher downtime and requiring more remedial workover operations, and results in higher operational expenditure.

Heavy oils in unconsolidated sandstones can be successfully developed with a nonthermal primary process called cold heavy oil production with sand (CHOPS) which can recover 5 percent to 10 percent of the PIIP. Paradoxically, this method requires the continuous production of sand into the well and uses a progressive cavity pump to recover the heavy oil and sand mixture to surface. The "sand cut" is initially up to 40 percent but over time this falls to a constant figure of less than 10 percent. The technique was developed in Canada, where it is widely used. It is also deployed in Colombia and in the former Soviet Union. The drawbacks include the vast amount of sand produced along with the heavy oil, which must be stored or transported and disposed of, leading to incremental Opex and environmental hazards associated with soil, air, and water pollution.

Where waxy, asphaltic crudes are being produced, the seller should be asked what remediation measures have been tried, such as hot solvent washes and cutting with slickline tools, what they cost, whether they have been effective, and how long before a repeat remediation is needed. The seller should be quizzed about whether there is a need for the infield pipelines to be insulated and heated, and whether extra chemicals are needed to "keep the crude flowing," such as spiking pour point depressant into the crude prior to it entering the export pipeline. Any extra costs or downtime incurred due to waxy crude production should be recorded as this will affect the economic modeling of the asset.

Any artificial lift methods employed in the field, such as sucker rod or beam pumps (i.e., nodding donkeys), electric submersible pumps, jet pumps, and gas lift, which are used to boost low reservoir energy to help lift liquids from the wells, should be reviewed, with emphasis on appropriateness, source, and cost of energy used, downtime history, and whether there are any untried methods that may improve field performance.

### 6.10.3 Gas production operations

Gas well productivity decreases as the reservoir pressure depletes, until the well ceases to flow. However, the presence of liquids, either coproduced with gas directly into the wellbore or condensed from vapor in the upper portion of the tubing, will accumulate in the well if the gas velocity is too low to remove them. This will cause the gas well deliverability to fall more rapidly. The phenomenon, known as liquid loading, causes gas to flow intermittently, eventually killing the well, and is recognized as a major challenge for gas producers worldwide. The PDR production engineer must determine whether the seller's gas producers are beset by liquid loading, and what steps are being taken by the seller to tackle the problem, as if it is identified and reduced, higher gas production rates are possible.

Recognizing the phenomenon is not always obvious, as a liquid loaded gas well can produce for a long time. But some tell-tale signs are sharp drops in rate and flowing temperature, and liquid slugs at the surface facilities, so plotting well flowrate and water gas ratio versus time and versus cumulative gas produced volume may highlight the onset of liquid loading. Remediation includes flowing the well at a higher velocity by installing a smaller diameter tubing (velocity string) or by creating a lower wellhead pressure (compression). Liquids can be pumped out of the well using various means, and foam can be injected into the impaired well to lower the density of the liquid column, allowing it to be to lifted from the well (Lea and Nickens, 2004).

One should ascertain what flow assurance measures are taken by the seller to maintain production if the reservoir temperature of the gas asset is relatively low. As produced gas cools on its way to surface, the likelihood increases that gas hydrates will form a plug in the flowline, which is particularly likely in offshore developments with long tieback flowlines on the seabed. The formation of a gas hydrate plug requires low temperature, high pressure; the presence of a hydrate former such as methane ($CH_4$), and sufficient quantities of water; all of which conditions exist in many producing gas wells.

The gas composition itself may be problematic, as gas heating quality is lowered as the percentage of inerts, nitrogen ($N_2$), and carbon dioxide ($CO_2$), increases, leading to lower gas prices and a lower income stream for the asset. If the export pipeline gas specification is not attained, then inerts removal will have to be installed at a major cost, or there will be a need to blend the gas with richer gas of lower inerts content from other fields sharing the export route.

If the gas contains a high percentage of the highly toxic component, hydrogen sulfide ($H_2S$), then stringent safety measures will have to be enforced. Its presence will also dictate which export routes to market are available as operators of sweet gas pipelines will not allow sour gas access. High percentages of $CO_2$ and $H_2S$ will require downhole corrosion inhibitor and/or specialist steels to be used for the well completions to make them more resistant to the corrosive nature of the produced gas, which means higher capital outlay. The priority data list for the M&A production engineer is shown in Table 6.7.

## 6.11 Role of the facilities engineer

When dealing with surface facilities it is vital to gather information on existing field production operations and infrastructure and surrounding terrain which will have a direct impact on what facilities may potentially be required in the event of a discovery. During the PDR visit, there is a need for the buyer's facilities engineer to familiarize himself with the assets' production system and its operating efficiency, the field layout, its location and environmental data, the condition and handling capacities of the

**Table 6.7** Production engineer's check list for material to gather in the data room.

| **Priority** | **Well completion data** |
|---|---|
| 1 | **Schematics of completion string with dimensions of casing and tubing.**<br>**Perforations (type and interval depths).**<br>**Workover results of acid washes and hydraulic fracture stimulations.**<br>**Field pressure surveys (preferably BHP), both static and flowing, with date and depth of measurement.** |
| 2 | Stimulation treatment data; schedules of slurry and rate; pre/posttest results. |
| *3* | *Daily well operations reports.*<br>*Completion fluid properties.* |
| ***Oil, gas, water production*** | |
| 1 | **Monthly production and injection records (by well, gathering centers, platforms, fields).**<br>**Field well tests of producers and injectors over time.**<br>**Field separator operating conditions over time.**<br>**Explanation of production allocation formula for tank batteries or gathering centers.** |
| 2 | Results of interference testing (if performed).<br>Results from observation wells (if available).<br>Results of interwell chemical tracer tests (if performed). |

facilities, the state of the infield flowlines, and the field export systems. The associated costs for drilling, construction, operations, transportation and, ultimately, abandonment for the assets on offer is a major priority for the M&A team's facilities engineer. It is useful to understand the approval process for field development and the typical timescale, including what level of detail needed in documents sent for approval, whether a safety case or an environmental footprint study is required, and whether there are any local content restrictions to comply with.

### 6.11.1 Field production and export facilities

The starting point of any facilities due diligence is to get acquainted with the field layout, with a general description of the facilities, including the main elements and their principal parameters, such as overall throughput, distance between elements, and type of export system. Also, it is useful to record the field location and any associated environmental data, as shown in Table 6.8.

**Table 6.8** Location and environmental data requirements for facilities due diligence.

| Onshore | Offshore |
|---|---|
| Location with reference to nearest towns | Latitude and longitude, distance to other fields |
| Height above sea level | Water depth |
| Local/surrounding terrain and topography | Metocean data; wind, waves, currents |
| Soil stability | Sea and air temperatures |
| Flooding risk | Hurricane risk |
| Subsidence risk | Subsidence risk |
| Earthquake zoning | Seabed conditions, including soils data |
| Nearest infrastructure | Nearest practical landfall |

The PDR facilities engineer must review the various process and utility challenges faced by the seller. Some of the issues listed below apply to each element of the field development; offshore, they will refer to wellhead platforms or Drilling-Production-Utilities-Quarters (DPUQ) platforms; onshore, they will refer to well sites, gathering centers, processing centers, and terminal facilities.

- Principal processing flowrates and pressures and a description of major items of hydrocarbon processing equipment, including produced water handling.
- Principal flowrates and pressures for seawater, compressed air, and produced water systems.
- Utility and electrical load lists; details of the main and emergency power generation systems.
- Weight for each principal element broken down into equipment, bulk, and steel.
- Process flow diagram, piping and instrumentation diagrams, and utility flow diagrams.
- Photographs of the site, of each piece of surface equipment, and of the adjacent area.

In addition to obtaining a diagram showing the flowline network, the following data should be gathered for each of the infield flowlines:

- Duty, such as three-phase production and end points.
- Location of the flowline within the overall network.
- Design parameters, in terms of flowrate, pressure, and temperature.
- Principal physical parameters, including diameter, length, coating system, and insulation.
- Whether the flowline is located on the surface, in a trench, or buried.
- What provisions are in place for pigging the flowline.

The seller should provide data on the existing regulations and plant performance with regard to produced oil-in-water specification and disposal route, gas flaring limits, and $H_2S$ and $SO_2$ discharge limits. In conjunction with these, data should be gathered on all export systems, including, but not limited to, those for the crude, gas, natural gas liquids (NGLs), and sulfur.

In the case of crude export, one would like a description of the system; the export rate, pressure and temperature; product specification (including true vapor pressure, Reid vapor pressure, and water content); storage volume; physical parameters of the export pipeline, such as diameter, length, elevation profile, and whether it is buried or not.

With respect to gas disposal, one would want to know whether the gas was processed prior to entering the export pipeline or whether it is reinjected; design parameters including flowrate, pressure, and temperature; product specification, Wobbe index, maximum $CO_2$, $H_2S$, and water contents; and the physical parameters for the pipeline as for crude.

The seller should be quizzed about whether he has plans for future expansion and how would this effect the facility capacity.

Finally, there should be a review of the present status of the asset's facilities, including any plans for refurbishment, and ultimately, abandonment. This may require a site visit by the M&A facilities engineer to "walk the lines" to ensure flowline integrity and signs of maintenance, to observe firsthand the state of the plant and equipment, and to witness the plant operators in action. The review should ascertain what is the age and condition of the plant, and what maintenance programs are in place and how closely are they adhered to by the seller. The review should consider what provision is made for corrosion monitoring, e.g., intelligent pigging of pipelines, and what refurbishment programs are underway or about to begin. The review should capture any abandonment and restoration requirements, including whether there are any decommissioned facilities requiring removal in the near term. A detailed and fully costed end of field life decommissioning plan should be requested from the seller.

### 6.11.2 Operating costs and production efficiency

As many assets are acquired with the operational team "intact," it is useful to review the asset's operational practices, to review the track record of the Operator (as the seller may be a partner in the operation), and to appreciate what operational difficulties have been experienced, and what areas might be improved upon. For example, what is the shift pattern used by the facilities staff and what is the current manning level? What is the split between company and contractor personnel, and if the asset is located overseas, what is the proportion of local to expatriate staff, and what is the perceived level of competence of the work force and what level of training is provided?

It is vital that the PDR facilities engineer obtains operating expenditure (Opex) data for input to the asset economic model. The annual fixed Opex is needed, along with a detailed breakdown on exactly what is included, such as personnel, logistics, maintenance and chemicals, support, insurance, and well intervention. Data are also required on variable Opex, such as tariffs for export, covering both transportation and processing, and details of any power import tariffs.

One vital piece of information to be gathered is the production efficiency (PE) of the assets, which can be defined as the total volume of petroleum produced annually as a percentage of either the economic maximum production potential or the maximum potential. Analysts claim that the former ratio gives a more accurate representation of PE as the export hubs of differing ages and at different stages of their producing life cycle become more comparable after uneconomic production potential is removed. If the PE of the seller's asset portfolio is falling, then this will be reflected in a faster natural depletion of the assets and reduce their EUR. For example, in 2018, the operators in the North Sea reported an average PE of 75%; a value they have managed to maintain for the past few years. However, less than half of UKCS Operators are currently meeting the PE targets set by the UK regulator of 80 to 85 percent and 90 to 95 percent for oil and gas installations, respectively. Once the PE has been determined, it can be used to adjust the buyer's due diligence production forecasts.

### 6.11.3 Capital costs

The facilities engineer is responsible for verifying the schedule for construction of capital expenditure (Capex). This will include the drilling program if no driller is present in the PDR, plus the abandonment costs (Abex). These will be provided to the economist for DCF modeling.

The field development schedule Capex should be provided in the data room and repeated in the IM. It is preferable if the total Capex is broken down into the costs for each field development element. In turn, each field development should be broken down into the major cost centers; an example of which is shown in Table 6.9 for a processing facility.

If a detailed Capex breakdown is not forthcoming from the seller, one should try to obtain cost norms for the country or region. These norms can

**Table 6.9** Major capital cost centers for an oil and gas process facility, onshore versus offshore.

| Onshore | Offshore |
|---|---|
| Drilling and completion | Well workovers and remediation |
| Equipment costs | Equipment costs |
| Bulk material costs | Bulk material costs |
| Fabrication and erection costs | Fabrication costs |
| Civil costs | Installation costs |
| Design and commissioning | Design and commissioning |
| Project management, cost and insurance | Project management and C&I |
| Abandonment | Abandonment |
| Contingency | Contingency |

**Table 6.10** Uncertainty in capex estimates versus stage of project.

| | Engineering/construction cost estimate certainty→ | | | |
|---|---|---|---|---|
| Quality class of estimate | Order of magnitude | Definition of project | Project AFE | Implementation AFE |
| 1 | +50%/−25% | | | |
| 2 | | +25%/−25% | | |
| 3 | | | +15%/−15% | |
| 4 | | | | +5%/−5% |

After Beliveau (2002).

take a number of formats including offshore platform topside cost in \$/tonne; steel jacket cost in \$/tonne; pipeline cost in \$/km-inch; design man-hours per tonne of topsides; and design man-hour cost. Capex estimates vary in quality depending on where one is in the life cycle of the development but should become more certain as the project matures as shown in Table 6.10.

## 6.11.4 Planning issues and governmental verification

If the assets on offer are green field projects, it is important to understand the process for approving a field development and what is the typical timescale. The buyer's facilities engineer needs to determine which authorities are involved, and if there is a particular agency with the ultimate "say so." Regarding the associated documentation, one must ascertain the level of detail required in the material submitted for approval. Also, whether a safety case is required, and what environmental studies are needed. Other governmental issues that should be investigated include whether there are any restrictions on the import of materials; whether any imports are subject to duty or other taxes; and if the government has to approve the chosen contractors. Potential buyers should also ascertain whether all oil produced must be sent to local refineries rather than be exported, and if so, what restrictions if any will apply. Further to the above, the buyer's facilities engineer should inquire about what certification and verification scheme is in place, and what companies are used for this activity, e.g., Lloyd's Register or Det Norske Veritas.

### 6.11.5 Local suppliers and contractors

If the opportunity is located in a new country or a region that is unfamiliar with the buyer, it is always advisable to question the seller about the materials suppliers and service contractors in his region of activity, as they can make or break an operation. With regard to materials, the buyer's facilities engineer should ask which companies are able to provide equipment, bulks, and steel, and whether manufacturing is done to internationally recognized standards. If not, buyers need to ask the seller what standards are applied. Concerning fabrication, determine which companies can fabricate and where are they located. On a nationwide scale, the buyer should ascertain the annual capacity and current utilization of fabrication facilities in the country, and ask what systems are in place to ensure quality. As to pipelines and installation, the buyer should find out which companies are involved in pipe laying and what capabilities they have, and which companies install facilities and what equipment is available for heavy lifts.

### 6.11.6 Infrastructure and refineries

It is crucial to obtain data on the available pipeline networks for transporting crude, gas, and NGLs in the region or country of interest. Data should include pipeline routing; pipeline length, diameter and design conditions; petroleum product specifications; condition; and refurbishment program.

In order to have a "Plan B," especially for onshore developments, the buyer's facilities engineer should ask the seller whether other transport routes exist, such as via road, river, canal, rail, and what is the relative cost and capacity of these alternatives, and what operational limits there are.

Specifically for foreign opportunities, the likely buyer must determine what facilities exist for transporting the crude oil out of the country,

**Table 6.11** Facilities engineer's check list for material to gather in the data room.

| Priority | Surface facilities, flowline constraints, operating environment |
|---|---|
| 1 | **Description of the field facility including overall capacity and layout.** **Design and operating parameters for each element (platform, gathering center) including flow rates and pressures.** **Supporting infrastructure including export systems.** **Capital costs including phasing.** **Operating costs.Production efficiency.** |
| 2 | National infrastructure including pipeline, rail, power, road systems. |
| 3 | *Local equipment suppliers and fabricators.* |

including the capacity (e.g., off-loading rate and capacity of storage facilities), size of tankers that can be handled, and whether these facilities are available for leasing. As a new entrant to the region, potential buyers must also inquire how do Operators get access to this infrastructure and what are the likely tariff arrangements. If the petroleum produced is to be sent for processing to the refineries in the country where the asset is located, one must determine the capacity of the existing refineries, their current status and condition, their typical feed stock composition, and what processes are involved.

If the production operations of the seller's assets require power from the local grid, the seller should provide the location of the main electricity grid, showing the proximity of transmission lines and substations to his assets, plus details of voltage levels, frequency, and capacity of the various sections. If the local power grid is accessible, can the seller purchase power and feed any excess power back, and what are the expected purchase and sales prices, respectively?

The priority data list for the facilities engineer is shown in Table 6.11.

## 6.12 Role of the economist

The M&A team economist must gather information on oil and gas sales prices in the region, existing infrastructure, E&P license terms, the region or country's overall commercial activity (especially in the energy sector), and ultimately develop an economic model to evaluate any existing producing opportunity or future discovery. Much of the regional background on fiscal regime, energy markets, role of the State in the energy sector, and views on political, financial, socioeconomic, and commercial risks associated with the country can be got from desktop research done prior to attending the data room.

A sound understanding of Capex levels is needed, usually derived from past schedules from existing assets, broken down into field development elements. Detailed Opex is vital, including any transportation and processing tariffs, plus details of the tariff formula. The seller should provide estimates of what it might cost currently to develop future fields, infrastructure, and export routes. The seller's future investment plans should be stated in the IM, given as inputs in the seller's economic model. The M&A economist must check that these agree with the approved working plan and budget (WP&B) that is usually detailed in the minutes of the various asset operational and technical meetings (OCM, TCM) between the asset Operators and their partners.

Regarding the contractual and fiscal terms for the particular assets on offer, PDR economists must ensure they are up to date and that they fully understand the mechanisms of the corresponding schemes.

For any license, one needs to know the details of the contracted parties; the effective date, duration, and expiry date of each license; whether an exploration program exists; the minimum work commitment; and if there are any outstanding work obligations. If there should be a future relinquishment of a license, ask if it can be a percentage of the contract area; what the critical dates are; and whether the partners in license have any preemption rights.

If the opportunity being evaluated is covered by a production sharing contract regime, one should ensure the following fiscal details are captured from the data room:

- bonuses, such as those associated with signature, a discovery, and start of production
- license area rental fees
- royalties, in cash or in kind
- cost recovery mechanism, via allowable expenditure, uplift, or recovery period
- maximum level of cost recovery in oil or gas
- government share of profit oil or gas
- income tax, including tax rate, depreciation mechanism, and any stability clauses

If the opportunity is covered by a concession regime, the flowing fiscal details should be captured:

- bonuses, such as that associated with signature
- license area rental fees
- royalties, in cash or in kind

- special petroleum tax, if applicable, including uplift and tax rate
- income tax, including tax rate and depreciation mechanism

For all fields, whether they are under development or on production, any tax data on costs incurred should be collected by the PDR economist. Also, as the commerciality of a project may hinge on it, the PDR economist should ask if operators have the right to market hydrocarbons (especially gas) in the country.

### 6.12.1 Petroleum price forecasts and sales contracts

The M&A economist must ensure that the seller has used the appropriate product prices for the indicative valuation of the asset portfolio given in the IM. If necessary, the assumed oil price forecasts should be adjusted to historical crude quality differentials with a recognized and appropriate marker crude. If the seller's produced oil shares the export pipeline with other crudes, the blend's average price should be reviewed, as a crude banking arrangement may exist to permit each crude to receive a price that reflects its quality. The price differentials between marker crudes are due to their relative volatility, sweetness, and transportation costs, where high API gravity and low sulfur and acidity along with proximity to market is more valuable.

Crude oil benchmarks, or marker crudes, were introduced in the 1980s to establish a standard for global trading. There are over 150 marker crudes in use today of which there are three widely used benchmarks against which other crudes and their futures contracts are priced; Brent is the European marker crude, Dubai (Fateh) is the traditional benchmark for crude produced in the Persian Gulf and destined to be sold in the Far East, and West Texas Intermediate is used in the United States. Other well-known blends include the OPEC Reference Basket (ORB), Malaysia's Tapis Crude which is traded in Asia, and Urals blend used for Russian exports. The ORB was introduced in 2005 to monitor global market movements and is currently made up of the price data of 14 crudes listed in Table 6.12.

Forecast gas prices will depend on the gross heating value of the sales gas composition after processing to remove liquids and/or nonhydrocarbon gases that the correct differentials are being applied. This processing will cause a shrinkage of the produced gas volume, as will using produced gas for fuel and flare requirements, and these volumes must be subtracted from the sales gas profiles.

**Table 6.12** Popular marker crudes for benchmarking oil prices around the world.

| | Marker crude | Country | °API | Sulfur wt% | Acidity mg KOH/g |
|---|---|---|---|---|---|
| Primary | Brent | UK | 40.2 | 0.35 | 0.06 |
| | Dubai (Fateh) | UAE | 31.0 | 2.00 | 0.15 |
| | WTI | USA | 39.6 | 0.24 | 0.09 |
| Other | Tapis | Malaysia | 42.7 | 0.04 | 0.22 |
| | Urals (to Med) | Russia | 31.7 | 1.35 | 0.07 |
| OPEC reference basket | Saharan blend | Algeria | 45.7 | 0.10 | 0.15 |
| | Girassol | Angola | 30.2 | 0.34 | 0.39 |
| | Djeno | Congo | 26.4 | 0.54 | 0.70 |
| | Oriente | Ecuador | 24.1 | 1.51 | 0.39 |
| | Zafiro | Equatorial Guinea | 30.0 | 0.25 | 0.77 |
| | Rabi light | Gabon | 33.2 | 0.14 | 0.08 |
| | Iran heavy | Iran | 29.6 | 2.24 | 0.15 |
| | Basra light | Iraq | 29.9 | 2.93 | 0.26 |
| | Kuwait export | Kuwait | 30.1 | 2.80 | 0.10 |
| | Es Sider | Libya | 36.7 | 0.37 | 0.10 |
| | Bonny light | Nigeria | 35.1 | 0.15 | 0.25 |
| | Qatar marine | Qatar | 32.7 | 1.85 | 0.70 |
| | Arab light | Saudi Arabia | 33.4 | 1.92 | 0.09 |
| | Murban | UAE | 40.5 | 0.74 | 0.05 |
| | Merey | Venezuela | 16.0 | 2.45 | 0.69 |

From various assay sources.

The M&A economist must ascertain whether there are specific gas sales agreements (GSA) for each asset that can include the pricing formula, the swing factor which is the seasonal difference in deliverability required, daily contract quotas (DCQ), annual contract quotas (ACQ), and delivery gas specifications for calorific value, water and liquids content, toxic components and inerts, and delivery pressure. All of these are needed to be captured in order to build a realistic economic model.

The PDR economist must become familiar with the GSA that has been negotiated for the assets under review. In general, GSAs are either depletion

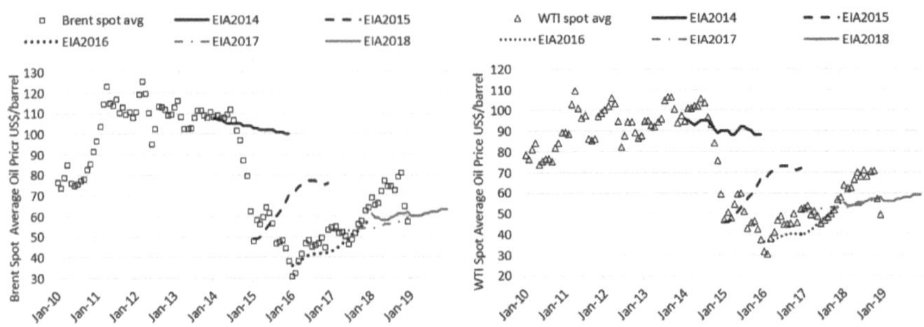


**Figure 6.9** Ever-changing forecasts for Brent and West Texas Intermediate crudes. *(After EIA, STEO archive 2014–18.)*

contracts or supply contracts. Depletion contracts are used for long-term gas supply, where producers commit to sell all production from a field to a buyer to meet its obligations. Supply contracts commit the producer to provide a set volume of gas for a fixed term, and it is the responsibility of the producer to source the gas, either from its own reserves or from third parties, to fulfill its obligations. One GSA element that is often worth investigating is the force majeure clause. Either party is rarely excused from failing to honor their respective obligations where market conditions have deteriorated to the extent that the GSA has become subcommercial. However, if the failure to supply gas is due to issues that the parties have no control over, then force majeure may be invoked to avoid financial penalties.

There are numerous forecasts of oil and gas prices provided by government departments, banks, sector analysts, and reserve auditing companies, but the only thing one can be sure of is that they will all invariably be wrong. An example of such forecasts for Brent and WTI prices from the US Energy Intelligence Administration (EIA) is shown in Fig. 6.9, which compares the actual marker crude prices with the short-term energy outlooks (STEO) for the past 5 years. It highlights the huge challenge in trying to predict the future oil prices.

## 6.12.2 Economic model

All sellers (or their financial advisors or bankers) have their own economic model of their assets. Buyers should strive to get hold a copy of this model (or as a last resort replicate it). By running the seller's model, the buyer can get an appreciation of what drives the profitability of the deal. The seller's model can then be adapted and changed into the buyer's own model as the buyer's own view of reserves and costs is firmed up. Using an economic model that

Table 6.13 Economist's checklist for material to gather in the data room.

| Priority | Commercial data |
|---|---|
| 1 | **Fixed/Variable costs or economic limit per well, for a group of wells, for the field**<br>**Cost of drilling and completing a typical well to each major subsurface horizon**<br>**Well Opex**<br>**Plant or Platform Opex**<br>**Transportation tariffs and Processing tariffs**<br>**Gas contractual obligations (quality specification, delivery pressure, swing, ACQ, DCQ)**<br>**Product prices forecasts, quality differentials**<br>**Crude assays (important if crude NOT sold on world market)** |

is acceptable to both parties is effectively a "common language" that will help to accelerate the deal by avoiding any commercial misunderstandings.

The priority data list for the M&A economist is shown in Table 6.13.

## 6.13 Role of the lawyer

Lawyers are not a common sight in PDRs, but buyers do call on them to advise on legal implications regarding assurance, tax, accounting, contractual liabilities, and potential unitization issues that may arise with any asset acquisition. They offer the same services to sellers, but from the divestiture perspective.

Although the individual responsibilities of each discipline are described above, it cannot be emphasized enough that those entering the fraught world of data rooms must remain skeptical. They must challenge the claims and assumptions made by the seller during their due diligence. Finally, M&A professionals must provide assurance to the buyer that they have captured the necessary information from the data room so that they sufficiently understand the downside of any deal.

The next chapter describes what must be done after the data room visit.

## CHAPTER 7

# After the data room

### Contents



Tired and footsore, the M&A team returns from its visit to the physical data room (PDR). Yet, there is likely to be no respite for the team members as their management (or client) will be eager to discover what they think about the assets about which they have gathered information. However, buyers must rein in their impatience for feedback from the data room team as there are some key tasks that need attention in order to ensure the due diligence process stays on schedule.

## 7.1 What to do after the data room

As one can imagine, the period just after a data room has closed tends to be one of intense interaction between the seller and his advisers and the buyers and their advisers. The seller is pushing buyers to complete their due diligence and submit a compliant bid by the time and date specified in the IM. The buyers are now under immense time pressure to meet the bid deadline and are desperately trying to assimilate the various views and opinions on the assets on offer to decide whether to pursue or drop the opportunity.

The nominated M&A team leader must continue to coordinate the buyer's team's efforts, including maintaining contact and continuing dialog with the seller and the managers of the data room, until the final report is ready for perusal by the buyer's management.


*Data Room Management for Mergers and Acquisitions in the Oil and Gas Industry*
ISSN 0376-7361
https://doi.org/10.1016/B978-0-444-63746-8.00007-7

### 7.1.1 Ongoing data management and evaluation

As before the data room, one member of the M&A team should be designated as "data manager" with the task of ensuring all data obtained during the PDR visit are uploaded onto the company database and merged with the downloaded data from the VDR. This needs to be done as soon as possible to allow its interrogation. By necessity, the post PDR analysis tends to be a "quick look" evaluation, which will be as rigorous as far as the time and data allow. The likely buyer must curb his enthusiasm to learn what is the view of the M&A team and allow sufficient time and allocate the necessary resources for further data loading and analyses after the PDR visit. Quality reports do take time to be written well, and they should be peer reviewed by the team leader, before being submitted to the buyer's management to aid their decision-making.

For good housekeeping, it is essential that the buyer's M&A team maintains complete and detailed records of what data were taken away from the data room. These original data should be kept "apart" from the buyer's own data and stored safely all together as they may be requested to be destroyed or returned afterward.

The buyer often makes copies of any key technical and commercial reports provided by the seller, but these copies should be numbered, and only made available to members of the M&A team. Buyers should not lose sight of the fact that these copies will most likely have to be destroyed too, so the buyer must keep a check on them and avoid any proliferation of them among the buyer's other staff.

### 7.1.2 Knowledge capture post-PDR

It is the individual responsibility of each member of the M&A team to write up their data room notes as soon as possible on return from the PDR. The team leader should collate these notes and share them with each member of the team. What tends to fall out of this transcript sharing exercise are the nagging questions that have yet to be asked or which were asked but have not yet been answered by the seller. It also reveals any remaining gaps in the dataset provided. These outcomes help when creating final requests for clarification from the seller on important issues, and for access to additional key data that were not in the data room.

### 7.1.3 Ongoing dialog between seller and buyers

The communication between the seller and potential buyers will continue up to the bid submission date. The management team running the data room on behalf of the seller tends to use a formal question and answer (Q&A) system that is usually incorporated within the VDR website, where attendees of the data room write down their questions and requests for extra data in the templates provided. The data room management team regularly compiles the questions and requests for data from each of the potential buyers and forwards them to the seller for a response. In most cases, the seller shares the Q&A list with every buyer as it creates a level playing field and saves having to answer the same question multiple times. The subsequent replies and any newly requested data are then uploaded into the VDR, so it is advisable for the nominated "data manager" of the buyer's M&A team to regularly log onto the VDR to check for any recent replies or additional data from the seller. Such information should be downloaded immediately and disseminated to the members of the M&A team, so keeping everyone in the loop and focused on delivering an evaluation report to their management by the bid deadline.

## 7.2 Traffic light confidence check on due diligence

While writing up the report, it is instructive for the M&A team to use a due diligence traffic light system to capture how well the data room exercise went. The system is a series of crucial questions to be asked or essential data to be requested to enable proper evaluation of each asset on offer, depending on the status of answers and data received, the asset will be assigned a traffic light color; red, amber, or green:

**Red** = Stop! Do not go ahead with the deal. There are simply too many unanswered questions, or swathes of essential data are missing, or the seller's reserve auditor work is lacking in quality. All of which suggests that far more due diligence is needed to satisfy the buyer.

**Amber** = Proceed with caution. There is some of the seller's work and the M&A team's own analytical studies in hand, but there could (and probably should) be more due diligence performed before the buyer is entirely comfortable with going ahead with the deal.

**Green** = Full steam ahead. It is apparent that sufficient due diligence has been undertaken and completed satisfactorily in accordance with the buyer's standards.

An example of the results of a traffic light process from a due diligence exercise of an asset portfolio in Colombia is given in Table 7.1. The portfolio is comprised of three producing fields and an exploration opportunity. The buyer has used Pareto analysis (Section 9.2) to rank the assets by perceived value, and they are listed in order of attractiveness, in columns from left to right, such that field C is preferred to field B, and so on. The exploration prospect X is the least attractive to the buyer.

**Table 7.1** Example of a traffic light grid, used to record the status of a due diligence exercise.

**Asset / Field Name** — Asset value/portfolio importance decreases to right as per Pareto Analysis!!

| Due Diligence Item | | Field C | Field B | Field A | Exploration Prospect X |
|---|---|---|---|---|---|
| COUNTRY RISK | Safe. stable business environment? | | | | |
| VENDOR RISK | Trust vendor – his claims? | | | | |
| DEAL RISK | Intent to do deal? Can we close? | | | | |
| SUBSURFACE | Regional Geology/data | Limited regional data integrated | Build better reservoir model? | Comparison w/ regional wells | Passable fit to proven plays? |
| | Geophysics/mapping | QC seismic/well tie?- is crest updip of well? | QC seismic/depth conv? Does ODT match closure? | Awaiting revised maps/QC seismic/well tie | Reliance ? |
| | Petrophysics | discovery well logs analysed | Re-evaluated Pilot Hole, ODT | Await answer MD/TVD discrep | Reliance ? |
| | Reservoir/geology | Proven disc./play type- 1 well | Re-evaluate n/gphi frac picture | Integrate regional data above | Proven play/analogues? |
| | Volumetrics/risk analysis | Completed but using Seller maps- await geophys review | Re-visit based on geophys? | Awaiting requested maps to finalise net pay/GRV inputs | Reliance ? |
| | Q/C Reserve Audit | Ext Auditor view unchanged | Auditor considered optimistic | Audit prior to App success | Reliance ? |
| | Production Hist. match/ DCA / MBAL | n/a | Unchanged- DCA based | n/a | n/a |
| | Well type, DD profile/EUR | Awaiting final MBAL model | Unchanged- DCA based | Update audit based on regional well | Reliance ? |
| | FDP concept | Need for 3D seismic pre drill? | Re-evaluate number of wells? | Review need for 3D seismic | Reliance ? |
| FACILITIES COSTS | Well/drilling costs | large risk on cost + timing | Update | Needs Review post wildcat | Reliance ? |
| | Facilities Costs | Update with revised well count | Update with revised well count | Needs Review post wildcat | Reliance ? |
| | Opex/transport costs | Can we export by barge? | Update | Needs Review post wildcat | Reliance ? |
| | Timing etc.? | CAN WE DRILL BY JULY 09? Else we lose license? Timing to approval devt EIA + permit? | Update | Needs Review post wildcat | Reliance ? |
| ECONOMICS | Economic model | Update as per above - Include stock options in model | | | |
| | Tax | Updated tax asset base | | | |
| | Financial Statements/Accounts | Update | | | |
| | Miscell. | Convertible notes update | | | |
| LEGAL / COMMERCIAL | Licence/Title/Expiry | Check title to area outside Blk | ANH excl. zones/blk. boundary | | |
| | Permitting etc. | Check permit drill/produce? | | | |
| | Crude Oil Sales Agreements | Check | | | |
| | Commitments/Work Programme etc. | Drill by July 09? Risk missing deadline? Cost of committed work program next phase? | | | |
| HSE | Regulatory Compliance | Check during field visit | | | |
| | Environmental | Check EIS completed/approved- national park issues | | | |

Each row of the traffic light grid is an item that requires due diligence, from general issues such as country risk and concerns about the seller's management team to detailed issues such as production history match, drilling costs, and commercial contract terms. Thus, each due diligence issue is reviewed for every asset and a corresponding color code of red, amber, or green is assigned.

If the subsequent traffic light grid shows too much red, then the buyer may decide to walk away, unless the seller can assure the buyer that his concerns are unfounded. In the example shown, there is a lot of red on field C, highlighting concerns over mapping, costs, and commercial arrangements. The exploration prospect X is believed to be extremely unreliable by the M&A team. The buyer decided to pass on this Colombian opportunity.

## 7.3 Submitting a bid or dropping out

The end game for the data room process is for potential buyers, after performing their due diligence, to be in a position where they can make an informed decision on whether to go ahead and submit a bid for the assets on offer, and what that bid should be.

Let us quickly summarize the data room workflow that has got the seller and the buyer to this point.

**a.** After compiling the material and data for due diligence of the assets to be sold, the seller canvassed potential buyers and asked those who expressed an interest in the deal to sign a confidentiality agreement and read the information memorandum.

**b.** Interested parties were then invited individually to a PDR and, prior to entry, the seller's management team presented its sales pitch on the opportunity to the potential buyers' teams. At the same time, permission was granted to potential buyers to access the seller's VDR and download any data therein for their due diligence purposes.

**c.** In the PDR, the M&A teams interrogated the seller's static and dynamic models, reviewed the seller's PIIP estimates and independently assessed past reservoir performance and verified production forecasts. To this end, the seller's ongoing/approved and future/unapproved development plans and associated costs were scrutinized.

**d.** In a mixed asset portfolio of exploration and production assets, the potential buyers would review the seller's prospect inventory, including future drilling targets and their associated risks, and comment on the potential incremental petroleum volumes that may stem from the proposed exploration program.

**e.** In addition to considering subsurface issues, the potential buyers assessed the integrity of the seller's operational practices, facilities, infrastructure, and any expansion plans for potential upside produced or injected volumes. While not always inspected via a site visit, the facilities were checked for short-term and long-term production constraints, along with associated technical, commercial, and environmental risks and possible mitigation plans.

**f.** Finally, the potential buyers ran economic models with inputs from independent forecasts of production, petroleum product prices, and cost projections. Besides being used to classify and characterize the seller's resources, the economic modeling provided appropriate valuations of the individual assets. Along with corresponding views of development and exploration risks, these valuations were ultimately fed through to the potential bid.

Aggregating the various valuations of reserves, contingent resources and prospective resources into a single bid number is not straightforward, as each resource and its corresponding development project bears different risks and challenges, and this is discussed in greater detail in Appendix 3.

If a bid can be formulated, the buyer must submit this by the deadline time and date, in the required format, and addressed to the assigned person responsible in the seller's team. These details are outlined in the IM and must not be forgotten in the race to perform the due diligence. It is worthwhile for the M&A team leader remind the team of the bid submission deadline to keep everyone focused on the deliverable.

If after performing due diligence the buyer decides not to proceed with a bid, and to drop out of the bidding process, then there will be a requirement to return or destroy the data that have been provided during the data room exercise.

## 7.4 Return or destruction of confidential data

One of the key provisions in the confidentiality agreement is that any proprietary data provided to a potential buyer must be returned or destroyed upon the request of the seller (Section 3.3.4). The buyer must confirm to the seller (in writing or via email) that the latter action has been done. Of course, there is always the risk that data will be misplaced, but such happenstances should not occur if the buyers are careful. By losing data, buyers lose face.

The previous chapters have considered the challenges in data gathering primarily for M&A deals involving conventional oil and gas resources. The next chapter describes instances when things go wrong.

# CHAPTER 8

# When things go pear shaped

Contents



There are many pitfalls one might encounter during a due diligence exercise of oil and gas properties in the confines of a data room and simply far too many to be covered in this book. This chapter describes almost two dozen examples (from the author's experience) that have occurred but may have been avoided if the approaches suggested in this book had been taken.

For the record, going pear shaped is English slang for when things go horribly wrong. The source of the phrase is purportedly the Royal Air Force in the 1940s, who referred to trainee pilots' failed attempts to loop the loop in their aircraft as "going pear shaped." Yet, mischievous wags


*Data Room Management for Mergers and Acquisitions in the Oil and Gas Industry*
ISSN 0376-7361
https://doi.org/10.1016/B978-0-444-63746-8.00008-9

claim it is a reference to a person's figure being narrower at the shoulders and wider at the hips, possibly due to the aging process and the effects of gravity.

## 8.1 VDR in Russian? не паникуйте!

The language used in the VDR is often one of the forgotten data formats, but it can be crucial.

Consider an M&A team that is given access to a VDR by a seller, who is a prominent oil and gas company that operates in the former Soviet Union. To their horror, the primarily English-speaking buyer's team discovered the VDR contained data that were in Cyrillic only. The M&A team had no idea what the names of the folders and the files they contained meant, and so had no idea where to start looking for key documents to be downloaded and translated. Due to the draconian timeframe imposed by the seller, it was feasible only to translate a small fraction of the data that were provided. Even then, it took the M&A team over a week just to get the VDR folder structure and the file names translated into English. To put this in context, a technical translator can usually translate 2000 words per day, which is around 8–10 pages of text. Of course, having one or two Russian speakers in the M&A team will help, but even they may be unable to identify all the key documents. This situation caused the planned duration of the due diligence to increase by over 50%.

In this instance, the challenge of a VDR in the Russian language had to be faced by every potential buyer's team, so each one of them had to download and translate the same documents, constituting a significant waste of time, money, and effort. In addition, there is always the risk that the various buyers' translations may be subtly different, leading to diverse meanings and understandings, so that different buyers may comprehend legal contracts differently, leading to variation in submitted bids, or no bids.

With some forethought, effort and expenditure from the seller and his advisers, the key data uploaded into the VDR could (and probably should) have been translated into English before allowing access to M&A teams.

## 8.2 2P or not 2P, that is the question

The Irish playwright, George Bernard Shaw, once said, "England and America are two countries separated by a common language," and this is certainly true of the different meanings attributed by the oil and gas industry to the term "2P reserves."

These 2P or proven plus probable reserves are the deterministically derived commercial volumes that M&A deals tend to be negotiated around, but very often the larger energy companies and sector analysts publish 2P reserves which are not compliant with the chosen resource reporting system. This criticism is less likely to be aimed at reserves auditors, who should know what they are talking about, and have a professional responsibility to present an unbiased and independent view to governments, regulatory agencies, and investors.

A known and respected energy sector analyst quotes 2P reserves in its regional and field studies, but when asked to clarify what it meant by "2P," it became clear that its estimate was not defined as per the PRMS. While deterministic 2P reserves are usually (incorrectly) assumed to be the same as probabilistic P50 reserves, the analyst admitted that its 2P value was closer to a riskier P35 value (Anderson, 2016). This is hardly surprising, as these energy sector analysts get their information from the field operators, whose numbers may be unaudited and who may be tempted to inflate them somewhat. In fact, most, if not all, of the major oil and gas companies submit reserves figures that are largely unaudited, which permits them to use their own in-house versions of the PRMS. This practice could be stopped, but it is up to a country's regulatory body to enforce strict compliance with its chosen resource reporting system. The United Kingdom claims to use PRMS for its national resource reporting, but many of the resource numbers submitted to the Oil and Gas Authority, the UK's regulator, are unaudited, which can lead to variable results for the same set of assets, as outlined in the following example.

In July 2008, the Abu Dhabi National Energy Company (Taqa Bratani) paid $631 million to acquire 100% interest and operatorship of seven mature oil fields in the northern North Sea from Shell UK Ltd. and Esso E&P Ltd. The deal also included 26.73% interest in the Hudson field, plus a 16% interest in the Brent pipeline system and a 24% interest in the Sullom Voe oil terminal. What was surprising was that Shell initially gave out two industry analyst reports on the assets to be sold to interested parties; no other data were provided. The stated 2P oil reserves for the seven fields are listed in Table 8.1, with two of the estimates coming from industry analyst

**Table 8.1** Comparison of "2P reserves" estimates from analysts for the "No Further Activity" scenario (ca. July 2007).

| Shell UKCS oil fields | Analyst A | Analyst B | Hyperbolic DCA |
|---|---|---|---|
| "2P" reserves (MMstb) | 91.6 | 81.0 | 48.8 |
| Difference to DCA | 88% | 66% | 0% |

reports and the third from a reserves auditor's hyperbolic decline curve analyses (DCA). For the latter, in the absence of any history matched field simulation models, DCA was performed on monthly field production reports in the public domain. It was assumed that the more severe exponential DCA would provide a 1P or proven reserves estimate, so the gentler hyperbolic DCA was deemed more appropriate to provide a 2P reserves estimate. The comparison of the various reserves estimates, assuming no further activity (NFA) by the operator, is astonishing, with the analysts forecasting two-thirds more to almost double that from rigorous DCA. It was discovered subsequently that the energy sector analysts' forecasts had included planned incremental recovery projects that were neither worked up sufficiently nor sanctioned which should not have been in the NFA forecast. This implies that the energy sector analysts' reported 2P reserves where actually a mixture of reserves and contingent resources, which is not permitted under PRMS′ guidance as the two classifications are subject to different risks and should not be added together (Offshore, 2011; Oil and Gas Authority, 2016).

Buyers should always ask the seller outright about the basis of the claimed 2P reserves figures and ascertain whether they are truly PRMS compliant, as it often transpires that they are not. Always bear in mind that energy sector analysts' reports are based on what operators tell them, so they may be optimistic in their outlook.

## 8.3 Reality check seller's regional exploration performance

Elvis Presley once said, "Truth is like the sun. You can shut it out for a time, but it ain't goin' away". Yet, this sound advice from the King of Rock and Roll is ignored time and time again when companies want to do a deal come what may, and refuse to face facts, especially if those facts paint the deal in a poor light.

An independent E&P company, based in the United Kingdom, hurriedly farmed into a reportedly massive deepwater opportunity in a microbasin of the Gulf of Mexico (GOM). The estimated cost of the exploration well was horrendous, but the potential prize was huge, according to the seller, who operated the lease. Unfortunately, for the UK company farming into the venture, the well was a dry hole. Questions were belatedly asked whether the wildcat should have been drilled in the first place as a postmortem of the farm-in suggested that the quoted prospective resources were far larger than anything previously been found in the microbasin. It transpired that if realistic potential in-place and recoverable volumes were used in the expected monetary value (EMV) calculation (see Section 9.1.1), the farm-in deal would likely never have been done. This tragic turn of events may have been avoided if the company farming into the prospect had performed some basic desktop research into the region, looking at past exploration well statistics for the GOM microbasin. Having gathered and synthesized wildcat data, one can use some simple techniques to confirm whether the seller's claims carry any weight. A couple of these tools are described below; the creaming curve and "missing fields" analysis.

### 8.3.1 Creaming curve

Examples of creaming curves are shown in Fig. 8.1, where the horizontal axis gives the number of wildcats in the order they were drilled.

With each discovery, the estimated resources are presented as cumulative values along the vertical axis. A steep creaming curve implies significant resources have been discovered with relatively few wells, whereas a shallow creaming curve implies the finds are small or many wells have been dry. As an example, let us look at the creaming curves for the Norwegian Continental Shelf in more detail.

In the period 2000−15, the Barents Sea had fewest wildcats, but substantial resources were found, as reflected in the steep creaming curve. Most discoveries in the Norwegian Sea were smaller than in the other seas, but its creaming curve shows a steady rise in accumulated resources. The North Sea shows limited resource growth for the early wells, then several major discoveries that steepen the creaming curve, then plateauing before making a huge vertical step with the Johan Sverdrup discovery, followed by moderate resource growth. Switching to the more recent period 2013−15, the Barents Sea clearly has the highest resource growth for the least number of wildcats. Resources found in the Norwegian Sea were roughly half those in the Barents Sea, yet more wildcats were drilled. The North Sea was on par with the Norwegian Sea for resource growth, but more than double the wells were required to prove the same volume, which reflects the relative maturity of the region (NPD, 2016a).

### 8.3.2 "Missing fields" analysis

The second technique uses field size distribution in the region of interest, as shown in Fig. 8.2. A statistical fit is made that honors the resource classes

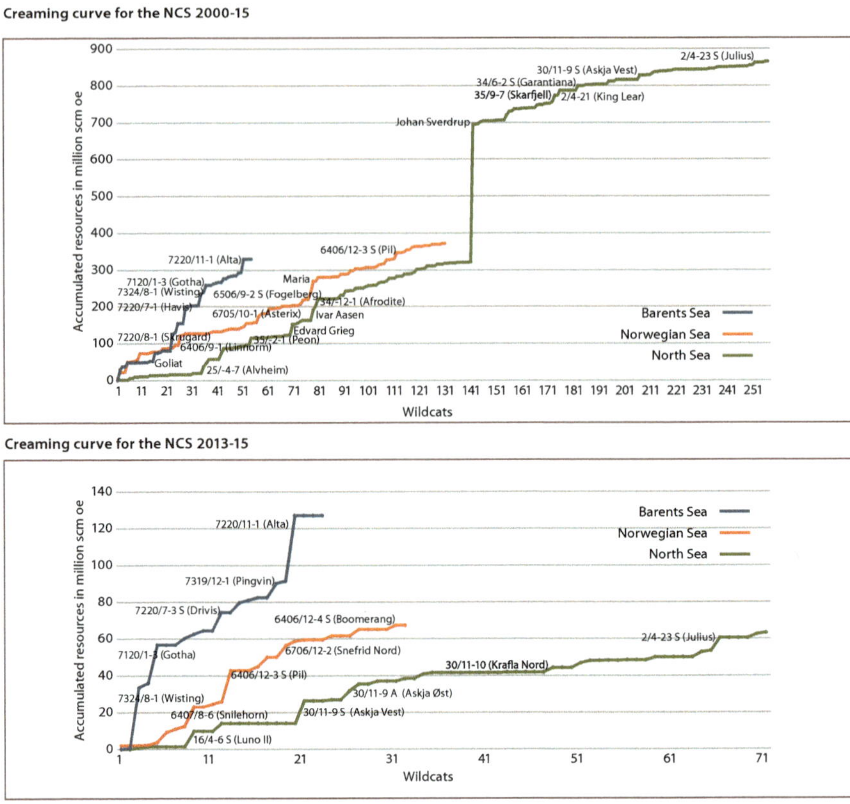


**Figure 8.1** Creaming curves showing the cumulative resource growth in the North, Norwegian, and Barents Seas between 2000 and 2015 and between 2013 and 2015. *(After NPD (2016a).)*

with the maximum number of fields (which appears to be a third-order polynomial equation in the example shown). The divergence from the statistical fit to the actual number of fields discovered gives the number of "missing fields" for each resource bin, which in the example is seven fields with potential recoverable resources of nearly one billion barrels of oil equivalent (see Appendix 2). The example shown cuts off the analysis (performed in 2004) at UK North Sea field sizes below 10 million barrels of oil equivalent (MMBOE), but this figure will be higher in the "lower oil price, for longer" environment that the industry must face over a decade later.

If the UK small independent E&P company had bothered to create a creaming curve and perform a "missing fields" analysis for the GOM microbasin of interest, it would have seen that it was very unlikely that the

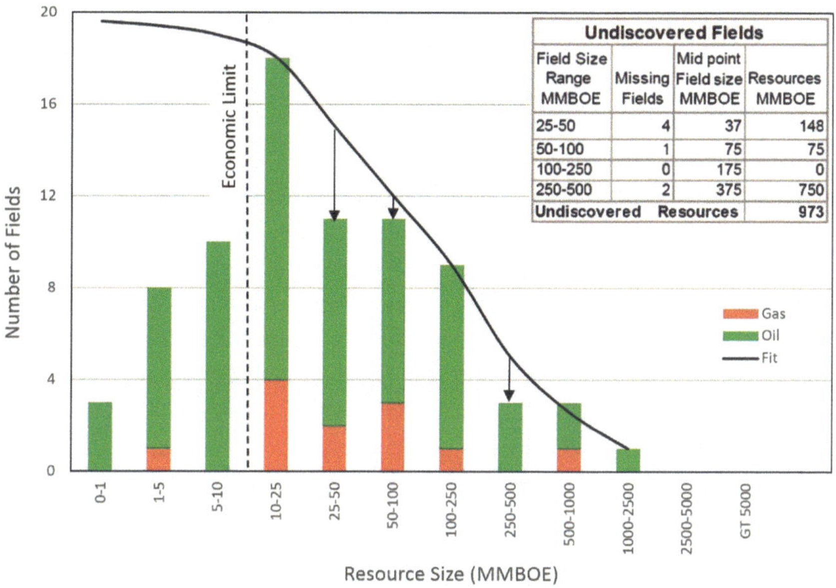


**Figure 8.2** Analysis of UKCS North Sea field size distribution implies upside resource potential of 973 MMBOE, identifying 7 "missing fields" in the ranges of 25–50, 50–100, and 250–500 MMBOE. *(After Rushworth (2004).)*

farm-in would have yielded such a huge discovery given the current knowledge of the region. However, that is not to say it might never happen, but the exploration team should have been made to much work harder to convince its management that the risk was worth taking.

## 8.4 Fudging hydrocarbon charge risk

The problem with explorers being allowed to take short cuts is that they often take the low road rather than the high road in order to get their prospects drilled.

Consider the following tale of an oil major trying to farm-out a working interest in an exploration license that was an ultradeepwater block in the southern Campos basin, offshore Brazil. Please note that there are several definitions of water depth in the oil and gas sector, but the more established criteria suggest deepwater is over 400 m, while ultradeepwater is over 1500 m ( API, 2011). A UK independent E&P company was determined to get into the region and sent an M&A team to the PDR in Rio de Janeiro to check out the farm-in opportunity. Assuming the Eocene play was proven, the M&A geologist estimated the prospect itself had a 60% chance that a reservoir existed, a 90% chance it was charged with hydrocarbons, but only a 50% chance that the reservoir was effectively sealed, which gave a geological chance of success (Pg) of 27%, as outlined below.

$$Pg = P(\text{trap}) \times P(\text{reservoir}) \times P(\text{hydrocarbon charge})$$
$$= 0.5 \times 0.6 \times 0.9 = 0.27.$$

Data mining revealed that discoveries adjacent to the farm-out block had much lower oil gravity values than the large producing fields further north in the Campos Basin. If the prospect contained such oil, it would severely decrease the price per barrel that one might obtain due to crude quality discounting and was likely to make any discovery subcommercial. However, the UK independent's exploration manager decided to evaluate the farm-in by assuming that, if a discovery was made, the gravity of the oil would likely be of average Campos quality of 25 degrees API. He argued that the uncertainty in oil quality could be handled by reducing the likelihood of hydrocarbon charge to 70 percent, which gave a lower geological chance of success of 21 percent, as shown below.

$$Pg = 0.5 \times 0.6 \times 0.7 = 0.21$$

This arbitrary change decreased slightly the expected monetary value (EMV) of the farm-in, but the risk weighted economic criterion remained positive and the opportunity was approved by the UK independent's senior management.

A better way to model the oil quality risk would have been to use a decision tree. Although the method takes longer to perform, it does capture better the inherent uncertainty of the farm-in with greater clarity and provides a more auditable answer.

From the decision tree shown in Fig. 8.3, using statistics from data mining, there is less than 10 percent chance of discovering higher quality oil required to make this project commercial. This is computed from adding the commercially successful elements of the decision tree, such that $0.00945 + 0.0567 + 0.02835 = 0.0945$ or a 9.45 percent chance. Yet, there is almost twice the likelihood, or a 17.55 percent chance, that any discovery will contain lower quality oil, which would be subcommercial in all cases. The subsequent EMV calculation from the decision tree is negative, and this farm-in may not have gained approval if this approach had been used.

However, the exploration well was drilled in the southern Campos Basin, and it did find a small accumulation of hydrocarbons, but the oil quality was less than 17 degrees API, which meant that the discovery was classified as subcommercial contingent resources under the PRMS; so much for taking the short cut of fudging the hydrocarbon charge risk!

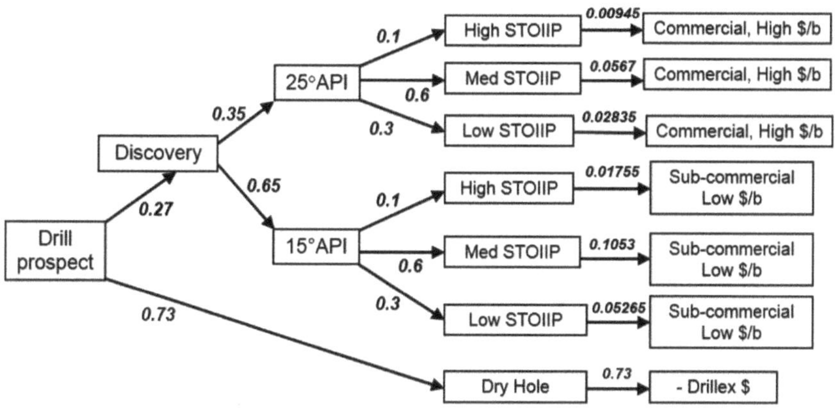


**Figure 8.3** Decision tree for assigning risk of which hydrocarbon type might be found in an undrilled prospect in the Southern Campos Basin, offshore Brazil. *(After Harrison (2012).)*

## 8.5 Assigning value to NGLs in an exploration farm-in

Some believe that overthinking creates problems that did not exist, but how should a company regard a farm-in opportunity where the prospective resources are broken down into oil, dry sales gas, and natural gas liquids (NGL) by the seller?

North Sea prospects are invariably assumed to contain volatile oil by sellers, who subsequently assign NGL volumes to the estimated PIIP. The practise arose as the price divide between oil and gas grew. Sellers argue that if an Operator can estimate the NGL yield from its processing of a produced gas stream with an "acceptable level of accuracy," then it should be included in the prospect evaluation, providing one remembers to include the incremental facilities cost of NGL recovery. Valuing exploration opportunities in this way results in the company farming into the venture being faced with a higher prospect valuation and receiving a lower working interest for the same investment amount.

Consider a farm-in opportunity in the North Sea, where the seller (who is also Operator of the license and an oil Major) estimates the 2U prospective resources in his prospect, which he assumes contains volatile oil, to be 18 MMstb of oil and 109 Bscf of wet gas; the latter is broken down into 91 Bscf of dry sales gas plus 9 MMstb of NGLs. (Investors should be aware that most oil Majors claim that fuel gas volumes are reserves too and only shrink the sales gas volume for losses due to flaring. Some regulators and auditors disagree with this practice.) The seller insists on the farm-in value being based on his resource estimate of oil, sales gas, and NGLs, which gives a higher value than if one evaluated the farm-in assuming just associated wet gas produced along with the oil. So, is there a problem?

One criticism of assigning value to NGLs in prospect evaluation of conventional resources is that seller does not know with any certainty the type of hydrocarbon present in an undrilled structure. Petroleum generation takes place as the breakdown of kerogen occurs with rising temperature over time. As formation temperature rises with progressive burial, immature kerogen is changed into oil and subsequently into wet gas by cracking and ultimately into dry gas, but the end of the oil generation window and the start gas generation window overlap as there is no clear distinction between them. Hence, mapping source rock potential across a basin will always have an inherent risk associated with the values prognosed.

Before estimating any hydrocarbon parameters for prospect evaluation, explorers should try and establish the likely hydrocarbon composition to classify the prospect in one of the six groups shown in Table 8.2. Note that earlier attempts at hydrocarbon type classification relied solely on API gravity, but this identifier is now regarded as unreliable, with better results obtained from the use of initial gas liquid ratio (GLRi) and heptane plus content (C7+) of the initial hydrocarbon composition (McCain et al., 2011). Once the hydrocarbon type has been classified, it follows that the method to compute PIIP is defined, as is the corresponding development scheme and estimated ultimate recovery (EUR). However, even if the reservoir hydrocarbon type is predicted correctly, the range of possible outcomes and spread of possible fluid properties for any of the hydrocarbon

**Table 8.2** Reliability of classification of hydrocarbon types using initial gas to liquids ratio, heptane plus content, and API oil gravity of the original in-place hydrocarbon composition.

| Reservoir hydrocarbon type | GLRi (scf/stb) | C7+ (mol %) | API gravity |
|---|---|---|---|
| *Reliability as an indicator* | *Good* | *Good* | *Poor* |
| Dry gas | >100,000 | <0.5 | 45+ |
| Wet gas | 15,000–100,000 | 0.5–4.0 | 45+ |
| Gas condensate | 3,200–15,000 | 4.0–12.9 | 45–60 |
| Volatile oil | 1,900–3,200 | 12.9–18.0 | 40–70 |
| Intermediate oil (from GLRi) | 1,500–1,900 | 18.0–26.5 | n/a |
| Black oil | <1,500 | >26.5 | 15–45 |

After McCain et al. (2011).

classifications make it a real challenge to forecast NGL yield with any degree of accuracy.

Returning to the farm-in example, there are enough data in the public domain that reveal changing oil properties with depth in the North Sea where the prospect is located, and to confirm many hydrocarbon discoveries in the prospect target zone, which was below 11,750 ft true vertical depth subsea (TVDSS). There have been many volatile oil discoveries in this zone, but there have also been a similar number of gas condensate finds. Hence, although the prospect may contain volatile oil, as suggested by the seller, it is equally as likely that the prospect may contain gas condensate.

Most oil and gas industry professionals believe that exploration companies should not attribute value to NGLs in prospect evaluation, arguing that the process is simply the first stage of a review to decide whether the venture is sufficiently interesting to justify the time and cost of due diligence. At this level, putting a value on NGLs is overscience, as the range of prospective resources calculated should adequately accommodate any additional benefit which may result from NGLs. Those with a facilities background argue that as NGL yield depends on the sum of the gas streams and relies on the export route available, one can easily have a variable NGL yield. The separation of NGL from either the oil or the gas streams is not an exact science, and the precise quantities in each export stream are dependent on the processing layout, the conditions employed, and the allowable specifications of each of the export schemes. Also, the assumed platform processing layout has a major impact on the well streams used in the analysis, as altering the separator pressure will result in different oil API gravity and changing choke settings may result in a different GOR values.

For the record, the seller did farm-out the prospect, the accumulation was drilled, and was found to contain gas condensate, with estimated contingent resources of 100 Bscf of wet gas and 4 MMstb of condensate. Over a decade later, the discovery remains undeveloped. The result was a major disappointment for the company farming in, who may have better spent its time trying to ascertain what was the chance of finding volatile oil.

## 8.6 Send the right people to the PDR

A seller was offering a farm-in opportunity into a deepwater discovery in the GOM and had arranged a time-limited PDR in Houston to allow interested buyers to access data to assess the deal. At the last minute, a buyer sent an inexperienced engineer to the PDR with the instruction to grab

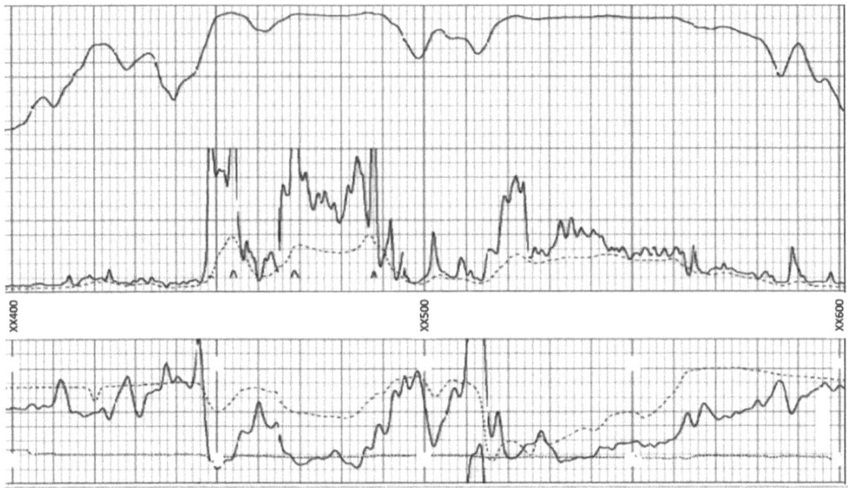


**Figure 8.4** Hurriedly photocopied wireline log prints from Gulf of Mexico discovery, taken from PDR, by a non-petrophysicist, with neither log header nor log codings and scales.

whatever data were available. The as-yet untested discovery well only had some log prints to take away. The engineer was not a petrophysicist, so before the PDR closed, he hurriedly photocopied the logs and returned to his HQ to present the data. An example of the fruits of his labors is shown in Fig. 8.4.

What the PDR novice had forgotten is that well logs are much more useful when the curves are presented with their scales and coding, so the log data he rushed to collect were almost useless.

The oft neglected log header can also provide details of hole size, casing dimensions, logging string make-up, tool types, and borehole stand-offs; all of which are needed for environmental corrections. Log headers also provide bottom hole temperatures over time since mud circulation was stopped, which can be used in a Horner plot to extrapolate to formation temperature, which is needed for pressure, volume, and temperature (PVT) correlations. Drilling mud properties are given, especially mud resistivities, which are essential input when analyzing spontaneous potential (SP) logs to estimate formation water salinity, which is a vital input to computing water saturation. Mud resistivity is needed to confirm that the appropriate deep sensing resistivity tool was run in the well; in general, laterologs should be run in salty muds, while induction tools in fresh muds and oil-based muds. Headers also give the logging engineer's comments, which chronicle any

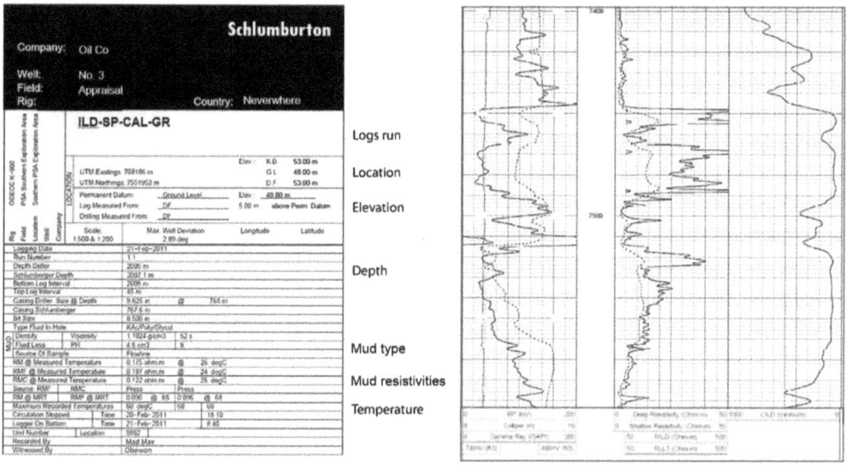


**Figure 8.5** Example of wireline log prints with full header information, scales, and coding of log curves.

issues with logging operations such as borehole washouts and permeable zones. A good example of wireline log prints with full header information, scales, and coding of log curves is shown in Fig. 8.5.

To conclude the example, there was no way to get back into the PDR, so the deal could not be evaluated, and the opportunity was missed.

## 8.7 Confirm the gas specification

An integrated energy company wanted to grow its upstream presence by acquiring a gas field portfolio of producing discovered and exploration assets and so develop a new core area for its business operations. The company sent a large multidisciplined M&A team to the PDR, but no-one in the team looked closely at gas quality of the discoveries as they were too busy checking gas volumes, costs, contracts, etc., of the producing fields. The resource volumes were confirmed via due diligence and the asset portfolio was purchased.

However, once the buyer began to better understand the assets, it realized that the discoveries, which lay toward the UK–Netherlands meridian and were the planned near-term future of the business, had gas compositions with high inert ($N_2$ and $CO_2$) contents and lower calorific values. This posed a serious problem as the export route's gas transmission specification would not permit access to the pipeline network if the produced gas stream had inert content over 7 percent. Unless enough sweet gas volumes from other producing fields were available to blend the inert content down to acceptable levels, the newcomer would have to pay a high financial penalty to sweeten its gas to gain access to the transmission system and so to the UK market. This may have been avoided if the buyer's team had included staff who had worked the area, as it is well documented that the inert content in gas increases as one moves from the UK waters toward the Dutch sector.

This sorry tale does not end there for the buyer. Along with the purchase of the portfolio of gas fields, the buyer was also offered the chance to buy into the operatorship of the existing export pipeline network. This infrastructure opportunity was eschewed by the buyer's management as being "non-strategic." However, when the buyer tried subsequently to bring its gas discoveries on stream, it found it had no leverage with the export pipeline owner-operators in the region. The buyer had to delay its development plans repeatedly, until eventually it was forced to build an expensive $CO_2$ removal plant onshore to get access to the export line and the UK grid to sell its gas. While this latter issue is not strictly a data room problem, it does reveal that buyers need to take a holistic view of any potential opportunity, as in this instance, the key contingency was access to the region's gas market.

## 8.8 Poor choice of analog

A farm-in opportunity arose in Central Europe, where the seller and Operator of a license claimed his tight gas discovery is an analog of a well-known basin centered gas accumulation (BCGA) in Wyoming, USA. The seller arranged a PDR and, during his management presentation, wheeled in expert after expert to support his claims. One of these experts used his own proprietary simulation software to produce many impressive production scenarios, all of which are stated to be grounded in the selected analog well performance.

The buyer was impressed, but cautious, and besides reviewing information from the data room, he instructed his M&A team to perform some background research, by accessing the claimed analog production data from the public domain and creating their own type well performance curves. New 3D seismic data acquired over the structure, coupled with updated geological studies, suggested that the accumulation, which lay in a basin that was much younger than the US basin, might not be such a cast iron BCGA after all. Also, the expert's simulated wells' performances were found to bear no resemblance to those generated from published analog well performances, being far too optimistic, which had a dramatic impact on the well numbers required and severely damaged the forecast project commerciality.

As the seller could not back up his stated numbers in the information memorandum (IM), the buyer withdrew from the deal. However, waiting in the wings was a US oil Major, who stepped in and subsequently spent $50 million on drilling several subcommercial wells before admitting defeat and exiting the venture and the country.

Just because a seller has a slick sales case and claims to have bespoke and proprietary databases and tools that support his forecasts, this does not guarantee a sound investment opportunity. Buyers must always bear in mind the phrase "caveat emptor" and perform their own due diligence. The seller in this example may still be seeking new paying partners, but do not be tempted!

## 8.9 Beware restricted access to production data

A seller in Latin America was only willing to release a couple of well files from his production database for due diligence by the buyer's M&A team. The well data provided showed excellent deliverability and performance, and the M&A reservoir engineers used DCA of the data from the two wells to create type wells in order to forecast recoverable oil volumes and well requirements in the field. The project economics looked good, so the deal went ahead. However, when the complete well database was made available, it revealed that the two wells provided in the PDR were the best two performers in the asset by far. Hence, the production forecasts on which the deal was based were hopelessly optimistic. Buyers must insist on being given access to all well production histories, not just those chosen by the seller. If the seller declines to do so, then the buyer cannot complete his due diligence, and should consider passing up the opportunity.

## 8.10 Insist on well check shot data

A seller sought a farm-in partner to drill a second appraisal well in an underappraised discovery located offshore North Africa. The target reservoir was complex and thin, and its structural relief was uncertain, so understanding the location of the oil water contact (OWC) was crucial. The seller claimed that of the two wells drilled in the structure, he only had access to one well with check shot data, which are used to calibrate the velocity versus depth model used in the subsequent seismic interpretation.

Note that seismic time interpretation provides a reservoir structure map, usually with interpreted faults, which is subsequently converted into a depth map using a velocity model. The conversion process multiplies the time map by a velocity map and is described in the equation:

$$\text{Depth (m)} = \text{Velocity (m/s)} \times \text{Time (s)}$$

However, the process of conversion to depth is subject to error, which can be expressed as follows:

$$\text{Depth error (m)} = \text{Velocity error (m/s)} \times \text{Time (s)} + \text{Time error (s)} \times \text{Velocity (m/s)}$$

With regard to the depth uncertainty in the structure map, the velocity error has far greater influence than the time error (which is also termed *pick error*), so the velocity must be constrained and calibrated by well check shot data (and vertical seismic profile (VSP) data if available).

Returning to the example, it is hugely significant that the seller claims that he cannot provide all the existing well check shot data to the buyer. Keen to do a deal, the buyer waived aside his M&A geophysicist's concerns and farmed into the venture. Once the deal was signed, the "missing" well check shot data were found, but when included in the seismic interpretation, they revealed the oil-bearing section of the reservoir may be only up to 30 m above the modeled OWC, which did not leave much room for error. The appraisal well was drilled and penetrated the crest of the reservoir some 25 m or so deeper than prognosed and the discovery of such a thin oil rim was written off as subcommercial.

M&A team geophysicists get frustrated when sellers provide depth-based geomodels in the PDR without either corresponding time-based models or velocity–depth relationships. If, after their review, they feel that the seller's PIIP ranges are too narrow, there is little they can do about it without having a full understanding of the seller's seismic interpretation. The M&A team must have access to all well velocity control data to complete its due diligence and buyers should consider walking away from any deal where essential seismic data fail to be provided by the seller.

## 8.11 Site visits are sometimes highly recommended

Site visits are costly and take time, both of which sellers and potential buyers do not like. However, if the deal is sufficiently large in volume and financial terms, and especially if it is a new country or in a new operating environment, then the buyer should consider a site visit by some of his M&A team.

Consider the example of a seller with mature onshore producing assets in the jungles of Colombia, Latin America, whose information memorandum (IM) stated that several gas discoveries that the seller had made were ripe for development, with all the necessary infrastructure in place, including infield tie-in lines, a processing plant, compression, and an export pipeline. However, the details of these assets provided in the data room were sketchy and the M&A team grew suspicious. The buyer prudently insisted on a site visit, and even went to the extent of hiring his own police escort and helicopter to fly down to the license in question. There was no sign of a plant or pipelines.

On his return to the United States, the buyer confronted the seller with his discovery, only to be told by the apologetic seller that there was "a misprint in the IM." Needless to say, the buyer did not go ahead with the deal.

## 8.12 Time gentlemen please

A potential buyer receives a small parcel of data from a seller, which allegedly includes some log data from an exploration well. The buyer is eager to see what the logs reveal and asks his M&A petrophysicist how long it will take to turn it around so that he can get back to the seller with a quick "are we interested, yes or no" reply. What time estimate should the log analysts give the buyer? A couple of hours? Half a day? Maybe a day or two?

Imagine the latter's shock when the package is opened to reveal a set of logs from the former Soviet Union.

A marked-up typical example of such log data is shown in Fig. 8.6, which shows the data being a poor photocopy of probably another photocopy, so there is no color coding of the logs. There is neither log header nor curve dash-dot coding, and the log scales are either absent or handwritten onto the log track in Cyrillic.

When analyzing log data such as these, one needs technical specialists and translators who can cope not only with the Cyrillic text but also understand Russian logging technology and wellsite operations.

As described in Section 8.1, these requirements increase project costs substantially, by thousands of dollars, and delay the analysis turnaround time significantly, by weeks in some cases. Buyers should not underestimate the extra time, effort, and investment needed to process data from the former Soviet Union.

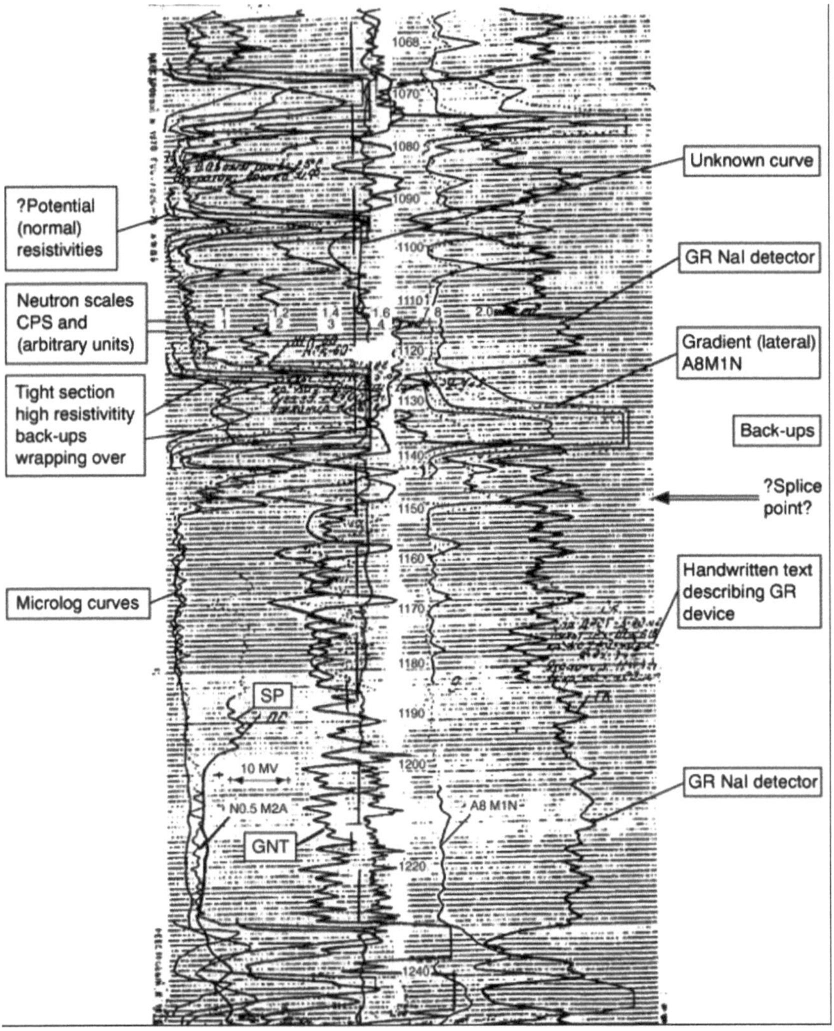


**Figure 8.6** Typical example of poor photocopy of wireline log prints from former Soviet Union, with neither header information nor coding of log curves, and only partial scales handwritten in Cyrillic in log tracks. This example has been marked-up by an expert in Russian logging operations.

## 8.13 Check fluid gradients all the way to the top

This example concerns a buyer who is considering farming into a large underappraised oil field in the Caspian Sea. The seller, who is a partner in the discovery, provided the potential buyer with some PVT parameters to characterize the undersaturated crude oil with a density of 36 degrees API, a gas oil ratio (GOR) of 900 scf/stb, and a bubble point pressure (Pbp) of 3699 psi, below which pressure then dissolved gas would be released from solution in the oil. The provided fluid measurements were made on a downhole oil sample taken from near the field OWC and were believed to be representative. Should the buyer be concerned?

The petroleum accumulation had a high structural relief of some 1000 m and was thought believed to have a reservoir pressure of 4640 psi. The buyer input the provided fluid and reservoir parameters into a PVT application, and the results are shown in Fig. 8.7.

The buyer became concerned that the reservoir may have a gas cap, as his pressure gradient calculation suggests the reservoir fluid may go below Pbp toward the crest of the reservoir structure. This would permit gas to escape from solution and create a gas cap. There was no local market for the gas, the host government had banned excessive flaring, and the estimate of petroleum initially in place (PIIP) currently being claimed by the seller was likely to be optimistic.

The buyer cross-checked concerns over the presence of gas with his geophysicist, who confirmed that the seismic data quality over the crest of the structure was relatively poor compared with that on the flanks, with loss of signal definition, which is often the case when a gas cap exists.

Unable to agree PIIP estimates with the seller, the buyer pulled out of the deal. As things turned out, the buyer chose the right course of action, as an up-dip appraisal well subsequently confirmed the presence of free gas in the structure.

## 8.14 Caution with well test press releases

Mark Twain, the celebrated American humorist and author, remarked that one should never let the truth get in the way of a good story. Investors should heed his advice and always treat media announcements of well flow rates with caution, as they can often be misleading.

For example, the reported flow rate may be too low as the test may have been restricted to minimize shut-in periods (to allow the well pressure to build up), or there may have been equipment restrictions such as small diameter test tubing in the well, or the use of a small test separator, or even flaring restrictions being imposed. In these cases, the Operator tends to state in the press release that the full well potential was not realized due to equipment constraints.

More worryingly for the investor is if the reported flow rate may be too high, which could have resulted from what is known as a "shareholders

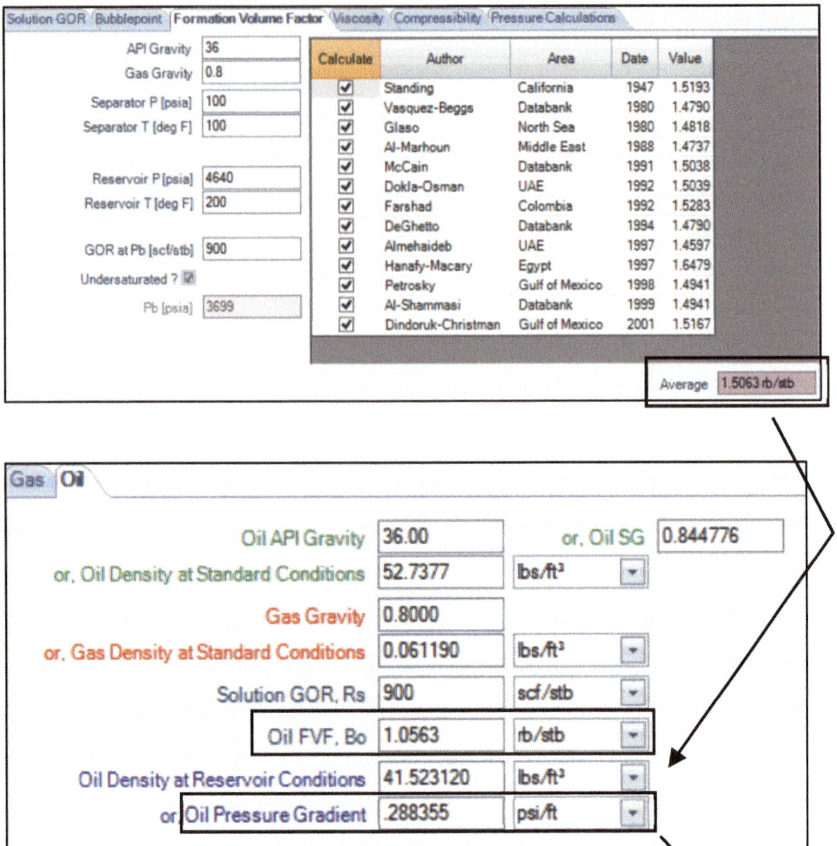


| | | | |
|---|---|---|---|
| a. | Bubble point pressure (from PVT) | | **3699 psi** |
| b. | Structural relief | | 1000 m |
| c. | Structural relief | (b x 3.281) | 3281 ft |
| d. | Oil gradient (from RE Toolkit) | | 0.29 psi/ft |
| e. | Relief pressure difference | (c x d) | 951 psi |
| f. | Initial reservoir pressure | (given) | 4640 psi |
| g. | Pressure at structure crest (f - e) | | **3689 psi** |
| **h.** | **Gas cap likely** (g < a) | | |

**Figure 8.7** Screenshots of oil formation volume factor estimate and in situ oil pressure gradient from Black Oil PVT. correlations, and corresponding calculation that suggests a gas cap may exist, as pressure at the crest of accumulation is lower than the oil bubble point pressure (Petroleum Solutions, 2018a) . *(With permission of Petroleum Solutions Ltd.)*

test," where all the commercial and subcommercial tested flow rates are added together and quoted in the press announcement. This practice is both misleading and incorrect.

One needs to take notice of which flow unit is used in the published statement, such as barrels per day versus barrels of oil per day; the former relates to the total fluid produced, be it oil, water, or whatever, whereas the latter refers to only oil.

Also, care needs to be exercised if the press release states that the published flow value is a "stabilized rate," as this could mean that unstable flow rates or associated water cuts have been "conveniently" omitted.

It is important to review all the circumstantial evidence, such as whether a very large choke size of 128/64 inches say, has been used during the flow test, as this may indicate that the published flow rate is "flat out" and is probably not representative of a sustainable rate.

One must always bear in mind that well flow tests do not normally give any indication of the long-term sustainability of the published test rate, so if high rates are stated, it is important to know whether the flow duration was a matter of minutes, hours, or days.

## 8.15 Reality check seller's exploration portfolio claims

The exploration portfolio shown in Fig. 8.8 is for the offshore blocks of Namibia on the western coast of Africa. The accompanying report extract

# Namibia

Once largely ignored by international oil and gas companies, Namibia has in recent times been called one of the last frontiers of oil and gas exploration. Despite currently having no proven oil reserves, it is believed that Namibia's offshore areas may hold significant hydrocarbon resources due to such areas being a part of the Equatorial Transform Margin, with geological similarities to prolific basins discovered in the pre-salt areas offshore Brazil.

Oil and gas exploration is in an early stage in Namibia, with only 17 wells drilled to date.

Freshfields Bruckhaus Deringer LLP Namibia March 2013

| Play | Prospect / Lead | Unrisked Prospective Oil Resources (MMbbl) | | |
|---|---|---|---|---|
| | | Low Estimate P90 | Best Technical Estimate P50 | High Estimate P10 |
| Upper Cretaceous Canyon-head Prospects | Prospect A | 170 | **374** | 684 |
| | **Prospect B** | 203 | **450** | 818 |
| | Prospect C | 482 | **1350** | 2913 |
| | Prospect D | 186 | **399** | 718 |
| | **Total** | | | |
| Upper Cretaceous Draped fan and channel Prospects | Prospect 4 | 58 | **192** | 467 |
| | Prospect 6 | 59 | **187** | 437 |
| | **Total** | | | |
| Upper Cretaceous Canyon-head Leads | Lead E | 280 | **702** | 1374 |
| | Lead G | 227 | **726** | 1615 |
| | Lead H | 85 | **266** | 579 |
| | **Total** | | | |
| Upper Cretaceous Draped fan and channel Leads | Lead 3 | 125 | **406** | 926 |
| | Lead 4 | 199 | **681** | 1578 |
| | Lead 5 | 98 | **346** | 783 |
| | **Total** | | | |

**Figure 8.8** Extract from report on exploration potential in Namibia's offshore areas and associated exploration portfolio. *(After FBP (2013).)*

claims the area is frontier and underappraised and ripe for exploration by international oil and gas companies.

However, investors should remember that oil and gas industry professionals suffer from overconfidence, which can manifest itself as a tendency to predict narrower ranges than their knowledge warrants. If one calculates the ratio of the P10 and P90 values in the table shown, it is clear that the portfolio has P10/P90 ratios in the range of four to eight. These values are more akin to step-out targets adjacent to producing fields, not frontier exploration, which tends to have a much wider P10/P90 ratio range of ten to fifty (Rose, 2001). If one assumes that resource estimates follow a lognormal distribution, which they invariably do, and assuming the high prospective resource estimates are structurally constrained, then the low estimates must surely have to be reduced, which in turn will reduce the best technical estimate.

## 8.16 Tendency to overestimate 2P EUR in the North Sea

Research has shown that many UK and Norwegian Operators overestimate their reported most likely estimated ultimate recovery (2P EUR) at sanction by around one-third (Hayes, 2016). Fig. 8.9 (*left*) shows the current 2P EUR post development drilling versus the predicted predrill figure for over 30 North Sea fields. Fig. 8.9 (*right*) shows how uncertainty at sanction can be reduced via more appraisal wells and so lower the 2P EUR ratio (i.e., 2P EUR current ÷ 2P EUR sanction), but the breakeven economic threshold of smaller discoveries in the North Sea precludes the necessary data gathering. This implies that small field developments in the North Sea,

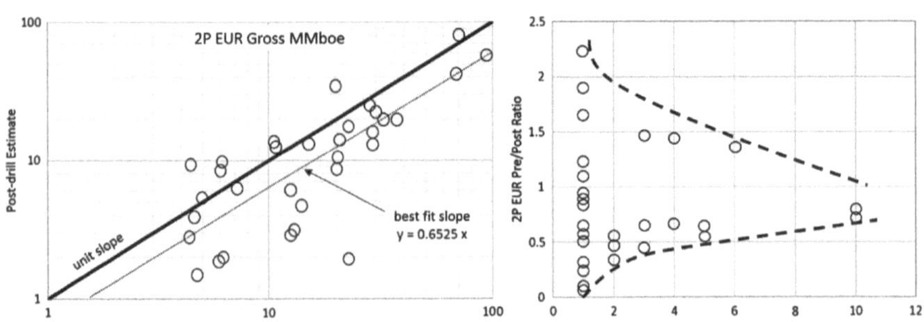


**Figure 8.9** Overestimation of 2P EUR is evident in the UKCS and Norwegian Continental Shelf; the effect is worse in smaller fields due to insufficient appraisal drilling being carried out. *(After Hayes (2016).)*

which make up the majority of recent and planned development activity in the region, tend to be underappraised and liable to fail to meet expectations.

It was found that small fields, defined as less than 50 MMBOE in the study, tended to get smaller. This is thought to be due to an industry-wide drive to meet economic targets leading to sanctioned reserves to creep above the actual P50 volume. In addition, small fields in the region are usually developed by subsea tieback to host production hubs, which means higher well intervention costs should workover be needed, and the claimed reserves are harder to recover. The corresponding reservoir models never really mature enough to be robust and the small well count means data are sparse and the field production forecast is invariably fragile as Operators cannot afford any well failures. Finally, the relatively short field life of such ventures means Operators usually only get one chance to get the development right. Potential buyers of such assets need to evaluate very carefully the seller's FDP.

## 8.17 Check PDR workstation capability with seller

Section 5.2 mentions the importance for a buyer to ascertain the number of workstations in the PDR, what data are loaded, and what software programs (with working licenses) are installed.

Yet, this is precisely what one potential buyer did not do, but went ahead and attended a PDR where the seller (an oil Major no less) proudly announced there would be six workstations available. This statement turned out to be true, but once in the PDR it was discovered that only two of the workstations had software that was enabled for use. Thus, many of the PDR team, expensively put together by the buyer, had to twiddle their thumbs waiting for their turn to interrogate the data on the enabled workstations. This state of affairs meant that less auditing work was done in the PDR and the subsequent due diligence after the PDR was put under greater time pressure.

## 8.18 Beware seller's dynamic simulation models without production history

A model can be defined as "a simplification of reality intended to promote understanding," but whether a model is correct is a value judgment, as it will only become evident over time whether a model is right or wrong. Hence, investors must always treat with caution any seller's forecast of recovery from an unconstrained dynamic simulation model, i.e., one which has limited or no production history to match to.

A seller was looking to raise capital and was offering to sell up to just less than half the working interest of his Siberian asset. The oil field in question had only recently been appraised and its development had just begun with several new producers being drilled and brought on stream. A PDR and VDR were set up by the seller, and interested parties were invited to visit the PDR in Mother Russia. The seller provided a dynamic model for inspection, which assumed significant sweep and pressure support from a large regional aquifer drive and forecast significant EUR from naturally flowing wells with simulated producing lifetimes of eight years or more.

Unfortunately, time constraints and visa requirements meant the likely buyer could not get his M&A team over to Moscow, so local consultants were hired to attend the PDR on the buyer's behalf. They checked the seller's dynamic model and collected screen dumps from the workstation and made accompanying notes which concluded that the model was "fit for purpose."

Back in London, the buyer's team became concerned as the recorded production forecast from the model in the PDR did not tally with the one reported in the seller's IM. Also, the model forecast recovery seemed optimistic given the reservoir was comprised of stacked fluvial sand channels with variable continuity and a review of published technical papers from the region failed to unearth any mention of large connected aquifers. In addition, the seller did not supply any analogs to corroborate the dynamic model assumptions.

The buyer was right to be concerned as, shortly after the PDR, some recent well production data from the Siberian field was uploaded to the VDR. The data showed the new wells coming on stream on natural flow, but dying in three to six months, showing lack of reservoir energy, and the wells having to have pumps installed to continue producing. This suggested the seller's dynamic simulation was inappropriate and that the provided FDP did not reflect the need for a pattern water flood, which would require many more production and injection wells. Ultimately, such a development would be far more costly than the seller would have potential investors believe.

This case study highlights not only the need to beware accepting at face value a production forecast from an unconstrained simulation model but also the problems that buyers face when trying to manage operations in another country with relatively unknown consultants whose technical capability and experience of due diligence may be lacking.

## 8.19 Check "automatic water shut-off" option in seller's dynamic model

Water handling in any producing field has a major influence over what recovery can be expected. Therefore, it is important to check what assumptions are made on how water breakthrough at producing wells is handled by the seller's dynamic model. More specifically, has automatic water shut-off been enabled in the simulation model, as selecting this option will permit the model to close off layers which have experienced water breakthrough and so and increase forecast recovery.

A buyer was struggling to reconcile the seller's higher EUR volumes which were derived from a history matched full field model of a large gas field, onshore Pakistan, that was underlain by a massive regional aquifer. The gas produced from the asset was dry with an initial CGR in single figures and the producing wells in the field had large diameter completions installed to achieve high initial gas rates to "outrun the encroaching aquifer." Having such large diameter tubing implied that wells would eventually load up with liquids and cease to flow at relatively high rates as aquifer advance was inevitable and unstoppable. However, the seller's model predicted the wells would continue to flow without liquid loading effects at lower rates than expected, and so recover more gas.

After a thorough review of the seller's dynamic model, it was discovered that the automatic water shut-off option was enabled in the simulator software. This meant that any producing layer, wherever it was located in well, would be switched off if water broke through during the simulation run. The model layers that were predicted to have water breakthrough occurred randomly throughout the entire reservoir sequence. Unfortunately, in field operations, it was only economically feasible to run plugs into wells to shut off water production from the lowermost layers in the reservoir. Neither this operational constraint was addressed in the seller's dynamic model nor were the number of workovers required to shut-of water producing layers in the wells accurately represented in the seller's economic model. Thus, the forecast gas recovery was too high and the economic model was missing future Opex which would result in too high estimated NPV and give an overly optimistic forecast of reserves.

After resolving the foibles in the model, there was much closer agreement between the EUR volumes of buyer and seller, and the deal continued to progress.

While not strictly data room catastrophes, the following two examples describe two production issues that commonly crop up; namely production well interference and poor choice of analog.

## 8.20 Drainage by older producers can affect future well recovery

If buyers have concerns about possible well interference between producing areas, they can use seller's data to perform DCA on a well by well level, and then plot the obtained EUR values for the wells according to the date they began production. This procedure usually highlights a trend in declining recovery by well, which is put down to each successive vintage of infill wells being drilled on a progressively smaller spacing, so that the EUR of the newer wells will be less due to the previous drainage by existing wells.

An example of this technique is shown in Fig. 8.10, which indicates that virtually all the latter wells in the gas field would likely recover less than 4 Bscf. However, the seller and Operator of the field had estimated (and booked) future EUR of 4 Bscf for each of the newer wells in the PUD locations using an outdated analogy, which failed to recognize drainage interference within the field. Once this error in judgment was highlighted, the seller had to write off several hundred Bscf of proved reserves for the project.

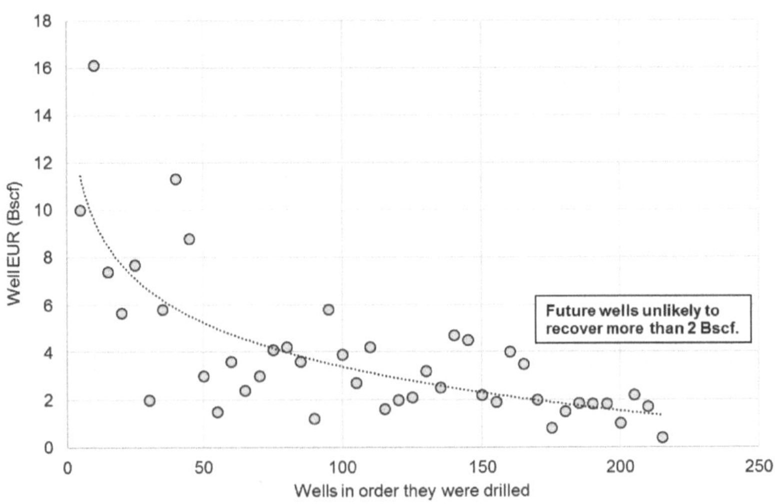


**Figure 8.10** Poor reserves estimation due to future well performance affected by drainage from older wells. *(After Harrell et al. (2004).)*

## 8.21 Poor analog selection can lead to reserves write down

The failure to select an appropriate analog can lead to a significant overestimate of reserves in an asset. In this example, the seller and Operator of a field assigned a minimum decline rate (Dmin) of 6 percent per annum to every producing well, unless the observed decline in a well was less than 6 percent, in which case the actual final stabilized decline was used. However, when the buyer plotted the well declines in the field, as shown in Fig. 8.11, it became clear that over half the wells had declines in excess of 12 percent per annum, and that barely 13 percent of the wells ever attained a value of Dmin of 6 percent or less. The buyer's reevaluation of the asset meant a substantial amount of proved developed reserves were written off.

## 8.22 Verifying high $CO_2$ content of a gas field when faced with limited conflicting data

An Antipodean oil and gas company was interested in acquiring a gas field in offshore Malaysia. Information from the data room visit suggested the field had high $CO_2$ content, yet the various sources, from downhole and surface, reported very different values of $CO_2$ content in the field, from 42 mol percent up to 92 mol percent (Table 8.3).

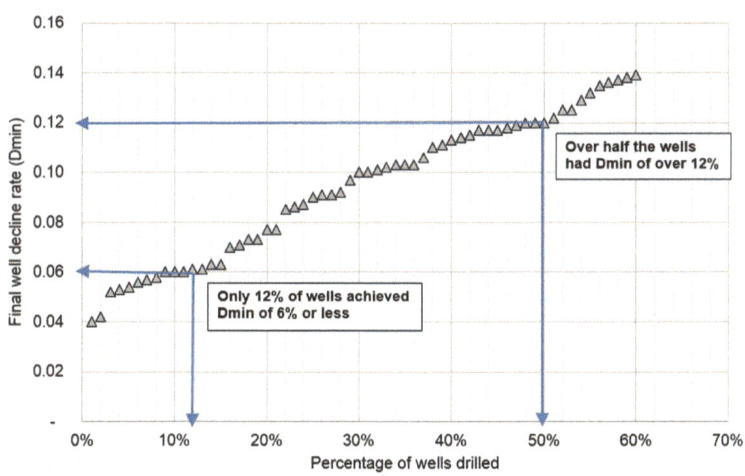


**Figure 8.11** Error in reserve estimation from inappropriate well performance analogy. *(After Ryder Scott (2004).)*

**Table 8.3** Variation in $CO_2$ content from different sources, wireline formation tester sample, drill stem test, and laboratory, for a gas field, offshore Malaysia.

| **Sample** | **$CO_2$ Composition (% mol)** |
|---|---|
| DST (on-site measurement) | 90 |
| DST (laboratory measurement) | 78 |
| WFT | 42 |

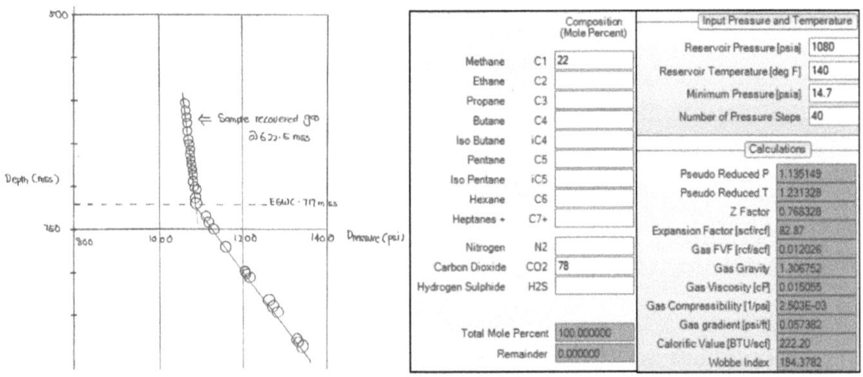


**Figure 8.12** Hand-drawn wireline pressure gradients and corresponding PVT analysis for dry gas with 78 mol percent $CO_2$ content for a gas field, offshore Malaysia (acknowledgment to Petroleum Solutions Ltd. software "*RE Toolkit*").

One must keep in mind that the Capex and Opex of high $CO_2$ developments are much higher than those of normal sweet gas fields. Separation methods are more suited to land-based operations, but operators in the region have improved membrane technology such that offshore fields with $CO_2$ content as high as 45 percent can be developed, reducing $CO_2$ content of exported gas to between 8 percent and 23 percent. However, developing gas fields containing over 60 percent $CO_2$ content is probably beyond the current economic limit (Isa and Azhar, 2009). Hence, it was imperative to determine the representativeness of the different PVT samples to validate the reported $CO_2$ content.

The approach was taken that as liquid content of produced gases in the region was very low, this implied that gas density was primarily a function of $CO_2$ content. This was verified from comparing the PVT-derived gas pressure gradient versus that from the wireline formation tester (WFT) measurements (Fig. 8.12), which indicated the surface samples gave more reliable $CO_2$ figure of 78 mol percent. This conclusion meant the company did not go ahead with this subcommercial acquisition. Note that the plot of WFT pressure verses vertical depth is hand drawn because the data room was located in Malaysia, where there are restrictions on what equipment can be taken into a data room and what (if any) data can be to taken away.

CHAPTER 9

# Tips for quicker focused evaluation

Contents



Arguably the biggest challenge in any data room exercise is keeping the due diligence on track while coping with the immense time pressure. This chapter offers some advice on how to ensure the due diligence goes according to plan. Some techniques are represented that can help keep the M&A team focused on processing sufficient data to allow it to provide the buyer with enough comfort and understanding about the deal to either walk away or to make an informed offer.


*Data Room Management for Mergers and Acquisitions in the Oil and Gas Industry*
ISSN 0376-7361
https://doi.org/10.1016/B978-0-444-63746-8.00009-0

## 9.1 EMV as a straight line for rapid farming analysis

The acquisition and disposal of exploration acreage is achieved via the mechanism of the farm-in and the farm-out, respectively. Farm-outs were introduced by the smaller independent oil and gas companies who wanted to gain a foothold in the large tranches of relatively unexplored acreage held by the larger oil Majors. The latter tend to have too much acreage with respect to their available internal funds for exploration, appraisal, and development. During spells of low oil prices, exploration budgets are squeezed and, to drill up commitment wells or lose the acreage, the oil Majors are forced to farm-out.

Farm-outs can be thought of as an indirect method of funding exploration as the operator of the acreage reduces or eliminates his dry hole risk in exchange for some of his equity in the acreage. The company farming in pays some or all the wildcat costs in anticipation that the well will be a commercial discovery and he will gain future production revenues from his newly acquired share in the acreage.

The following section explains how the expected monetary value (EMV) and exploration risk assessment methods can be applied to evaluate potential farm-in and farm-out deals. More importantly, the section describes how EMV can be expressed as a straight line for rapid deal evaluation.

### 9.1.1 Definition of EMV

Consider the decision tree shown in Fig. 9.1 which depicts two possible outcomes when drilling an exploration well: oil discovery or dry hole. The values of each outcome are a positive net present value (NPV) and a rig cost (−D), for the discovery and dry hole, respectively. The probability of finding oil is denoted by "Pg," and it follows that the probability of drilling a dry hole is therefore equal to (1−Pg).

The EMV of the above scenario is simply the sum of probability weighted outcomes.

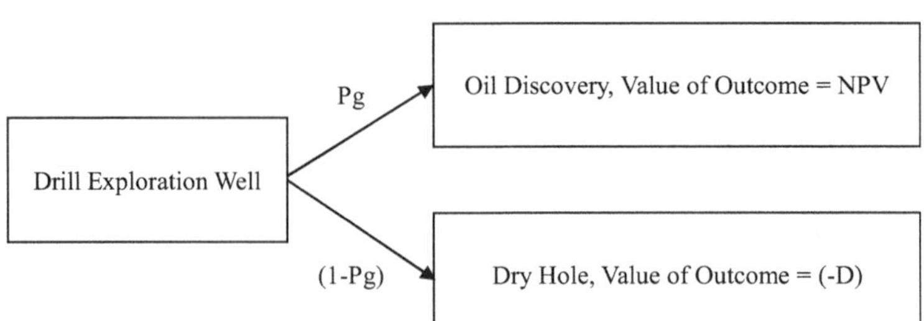


**Figure 9.1** Decision tree for an exploration well with two outcomes.

$$\begin{aligned}\mathbf{EMV} &= \text{Pg} \times (\text{NPV}) + (1 - \text{Pg}) \times (-\text{D}) \\ &= \text{Pg} \times (\text{NPV}) - \text{D} + \text{Pg} \times (\text{D}) \\ &= \mathbf{Pg} \times (\mathbf{NPV} + \mathbf{D}) - \mathbf{D}\end{aligned}$$

The last expression shown above takes the form "y = m (x) + c," which implies that the EMV of an exploration well can be characterized by law of a straight line (Antia, 1988). It is clear that a key driver of EMV is the cost of drilling the exploration well. If an oil and gas company is a low-cost operator, it can be more competitive than its rivals. For example, being able to drill cheaper than its rivals, it can accept riskier prospects of the same value and it can drill smaller, more marginal prospects of the same risk.

The graphical form of the straight line EMV equation is shown in Fig. 9.2. From the above relationship, the minimum exploration risk (Pmin) is the lowest chance of success allowable for the given project value and exploration well cost. The application of this linear expression of EMV will be shown in the next sections on farming analysis.

### 9.1.2 Farm-in analysis using straight line EMV

Motives for farming into acreage can range from an oil and gas company having funds to explore, but no suitable acreage to drill, or fostered in the belief that it has better knowledge and expertise than the acreage holder, or by the desire either to operate or to learn from a more experienced

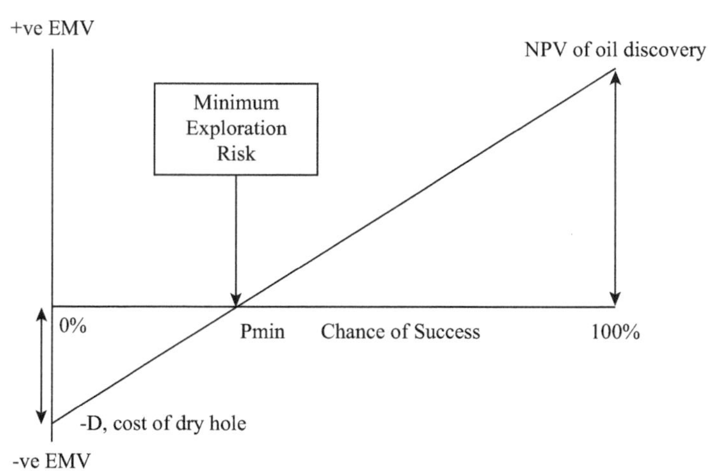


**Figure 9.2** Two outcome expected monetary value (EMV) of an exploration well shown as a straight line.

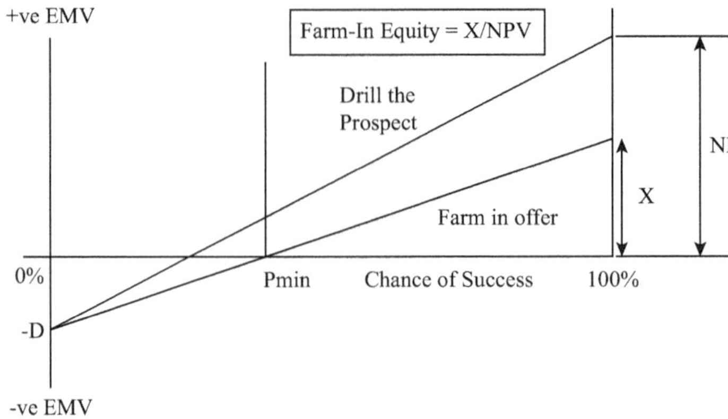


**Figure 9.3** Farm-in analysis using straight line expected monetary value (EMV). *(From a buyer's perspective.)*

Operator. The oil and gas company wanting to farm-in may also be in a tax position whereby it can carry out exploratory drilling more cheaply, which can arise when a country's tax regime allows company's paying tax to take equity in an exploration program and offset some of the cost against their tax bill. The evaluation of a farm-in requires the determination of the EMV of the prospect derived at a range of different success probabilities from the buyer's point of view. These are plotted as shown on Fig. 9.3, which shows that the acquiring company will farm-in to the acreage for an interest of "X ÷ NPV" in exchange for taking on a commitment to drill the exploration well. The buyer will incur dry hole cost, -D. The farm-in deal will only go ahead if the buyer's perceived chance of success for the specific prospect in question is greater than probability, Pmin.

### 9.1.3 Farm-out analysis using straight line EMV

The reasons for farming-out acreage can be many and varied. Perhaps the seller lacks money or time to fulfill the work obligation on the acreage or lacks confidence in the potential of the acreage. It is possible that the seller may not have the expertise to evaluate the acreage or that disagreements with partners are preventing appraisal activity. It is also possible that the seller owns infrastructure in the area and is seeking to tariff third parties. The evaluation requires the determination of the EMV of the prospect derived at a range of different success probabilities from the seller's point of view. These are plotted in Fig. 9.4, which shows how the seller will try to farm-out an interest of "Y ÷ NPV" of the acreage in exchange for a commitment to drill the exploration well. The acquiring company will incur the dry hole cost, −D.

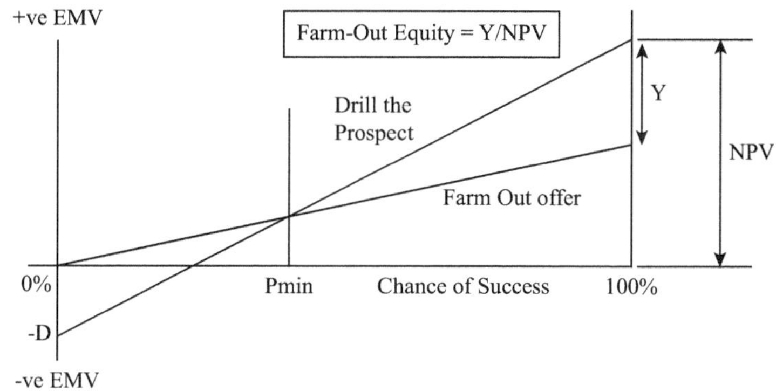


**Figure 9.4** Farm-out analysis using straight line expected monetary value (EMV). *(From a seller's perspective.)*

The farm-out will only go ahead if the seller truly believes that the minimum exploration risk for the specific prospect in question is less than probability, Pmin. If the perceived chance of success is greater than Pmin, then the seller should drill the prospect himself as he will give away value by farming out.

There are other plots that incorporate EMV which can be used to focus one's efforts when evaluating a prospect inventory, and these are described in the next sections.

## 9.2 Pareto analysis for ranking assets

Pareto Analysis is a statistical technique in decision-making used for the selection of a limited number of tasks that produce significant overall effect. It takes its name from the 19th century Italian social economist, Vilfredo Pareto, who observed that 80 percent of income in Italy went to 20 pecent of the population, and this finding became known as the 80/20 rule. He later carried out surveys in other European countries and found that a similar distribution applied. Its underlying principle is that by doing 20 percent of the essential work, one can generate 80 percent of the benefit of doing the entire job, which is referred to as "the vital few and the trivial many."

The 80/20 rule can be applied to almost anything, but it does treat all elements under investigation the same. Hence, when faced with evaluating a sizable asset portfolio of producing fields, discoveries, and prospects, one should break out the different assets into their relative stages of maturity before carrying out the Pareto ranking exercise. Similarly, when faced with a mixture of predominantly oil producing fields and mainly gas producing

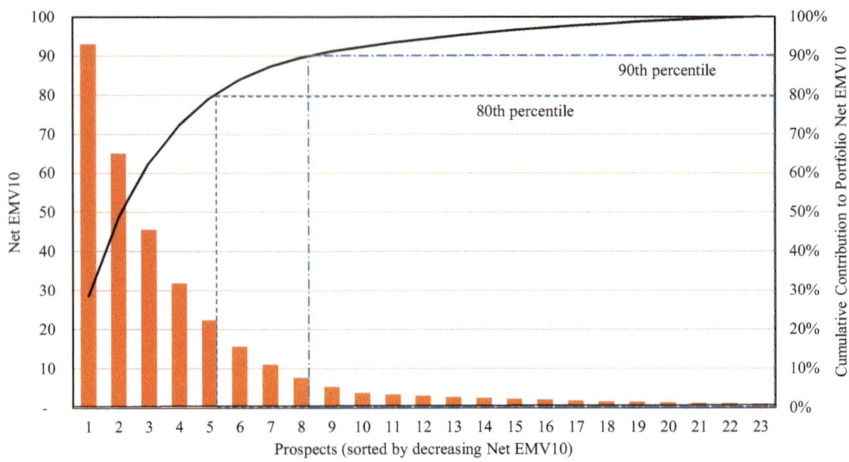


**Figure 9.5** Graphical representation of a Pareto analysis of a prospect portfolio.

fields, the ranking parameter employed in the Pareto analysis should be value-based as opposed to volume-based, i.e., avoid the use of the misleading barrels of oil equivalent measure for the reasons outlined in Appendix 2.

Let us consider an exploration asset portfolio which has been analyzed for volume and value, and then risked, to reflect the chance of geological success. The selected ranking parameter of the subsequent due diligence results of the portfolio is then sorted in ascending order and plotted as a cumulative contribution curve as shown in Fig. 9.5.

The exploration prospects in the example are sorted by the seller's net entitlement, expressed as expected monetary value at a 10 percent discount rate (EMV10) to a chosen effective date. The Pareto analysis highlights that 80 percent of the portfolio's risked value is held in only five of the assets; these are the ones on which buyers should focus their efforts.

## 9.3 Tri-plot for ranking assets

Another useful tool for focusing one's efforts in a due diligence exercise is the tri-plot.

The example in Fig. 9.6 is for an exploration asset portfolio, but the technique can also be used for contingent resources using a chance of development (Pd) in place of chance of geological success (Pg). The tri-plot allows one to display volume, value, and risk for each asset that is under

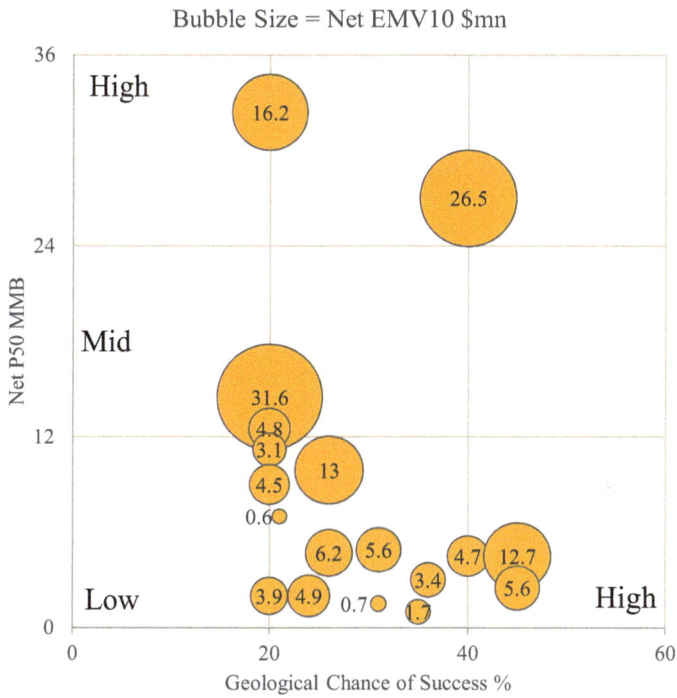


**Figure 9.6** The Tri-plot permits the estimated volume, value, and risk of a portfolio to be represented on a single plot (Harrison, 2012).

scrutiny. The axes are often rescaled to position the asset portfolio data into the low, medium, high categories to allow one to visualize more clearly which assets may be relatively more important to the buyer.

## 9.4 Other EMV-based plots for ranking exploration portfolios

An E&P company will strive to maximize the EMV of its exploration portfolio for the smallest capital outlay. To achieve this goal, those prospects that have a high EMV net to the company, but with low net work commitment costs, will be drilled first. Those prospects in the portfolio that have low net EMV to the company and which have high net work commitments should be farmed out. To identify which prospects should be drilled and which ones should be farmed out, one can use the plot shown in Fig. 9.7.

To optimize its exploration asset portfolio, the E&P company should identify and drill as many prospects which lie above the unit slope;

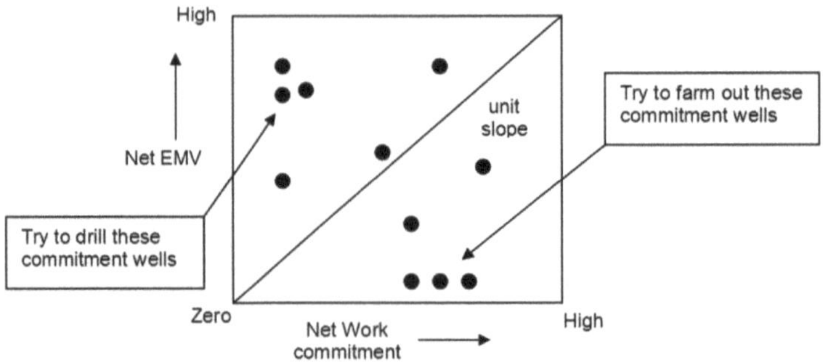


**Figure 9.7** Crossplot of net expected monetary value (EMV) and net work commitment identifies prospects to be farmed out.

prospects that lie below the unit slope should be aggressively farmed out at the earliest opportunity to lower commitment to these risky ventures. The company can also use the plot to identify prospects in which it should retain higher equity as these may provide more than $1 of value for every $1 spent on the agreed work program for the acreage.

A cruder version of the previous method is to plot the geological chance of success versus its corresponding EMV for the assets in the prospect inventory as shown in Fig. 9.8. Those prospects with a high EMV and a high geological chance of success should get priority when drilling budgets are reduced in times of low petroleum prices. The less attractive prospects, with lower geological chance of success and lower EMV, should be farmed out.

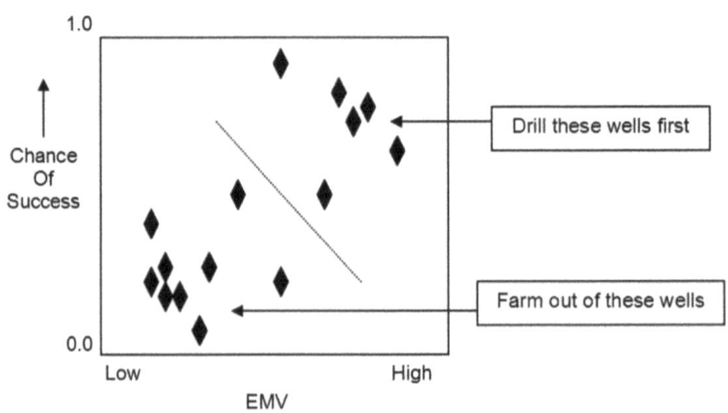


**Figure 9.8** Crossplot of geological chance of success and expected monetary value (EMV) identifies prospects to be farmed out.

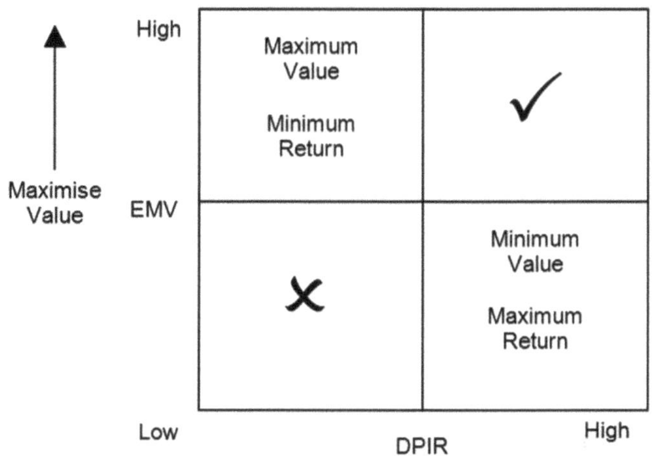


**Figure 9.9** Low–high matrix plot of expected monetary value and discounted profit-to-investment ratio to select an appropriate strategy for dealing with prospects in an exploration portfolio.

The third technique for maximizing value and return from an exploration portfolio is to plot EMV versus discounted profit-to-investment ratio (DPIR) in a matrix, as shown in Fig. 9.9, to highlight the high value—high return prospects to be drilled.

DPIR, which is also known as profitability index, is the ratio of profit to investment of a proposed project. It is calculated by dividing the sum of either the discounted net operating income or the discounted net cash flow from a project by the sum of the discounted investments. If net operating income is used, a DPIR of one is the breakeven value where the investment is just recovered; whereas if net cash flow is used, breakeven DPIR is zero. One should clearly state which definition has been used.

Companies often use DPIR as a project selection tool as it quantifies the amount of value created per unit of investment. When faced with limited capital, companies rank projects in decreasing order of DPIR and select projects from the list until the capital available for investment is used up. However, the technique is criticized for ignoring project risk when maximizing the EMV of a portfolio and will tend to produce a high-risk inventory.

The EMV-DPIR matrix shown in Fig. 9.9 suggests the following characteristics for prospects that are mapped in each of its quadrants:

- High value and low return—such prospects may have larger potential reserves, but they will need substantial capital expenditure to see them developed.

- Low value and high return—prospects with these characteristics may have smaller potential reserves, but they will need little outlay to bring them on stream. These projects could be relatively small satellite structures adjacent to existing oil and gas infrastructure, which could be tied back to a host platform relatively cheaply.
- Low value and low return—such prospects should be farmed out at the earliest opportunity.
- High value and high return—equity in such prospects should be retained or grown, and they should be drilled by the Operator.

## 9.5 Risk cover

Some explorers prefer to use risk cover when evaluating a prospect portfolio. This ranking metric is the EMV of the outcome of an exploration venture expressed as a ratio instead of positive or negative values. It is defined as the number of times an identical prospect can be drilled and the dry hole costs can be recovered from the discovery and the subsequent development of a single discovery. Consider the simple example of risk cover shown in Fig. 9.10, where a particular "ready to drill" prospect in a portfolio under scrutiny has a chance of one in eight of being a discovery.

$$\begin{aligned}\text{Risk Cover} &= \text{NPV of one successful development} \div \text{Dry Hole Cost}\\ &= 80 \div 10 = 8\end{aligned}$$

The buyer may prefer all prospects to be similar to the above example and choose to apply the calculated risk cover to the entire portfolio. By setting a target value of 8, the buyer then only considers prospects that exhibit risk cover that is greater than this value.

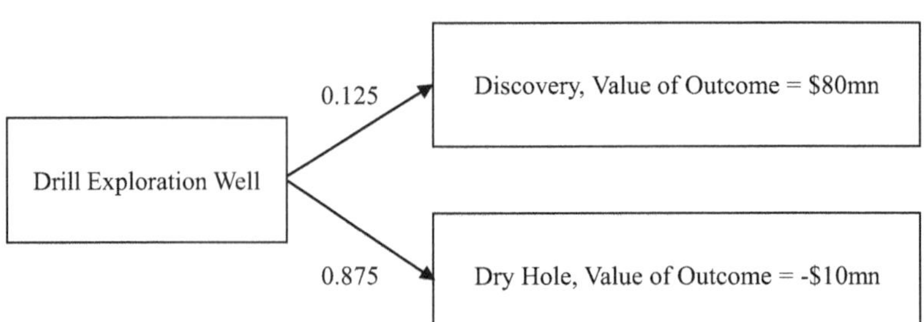


**Figure 9.10** Decision tree for a prospect with a risk cover of 8.

## 9.6 PVT checker

Potential buyers should always confirm that the quoted PVT parameters for the oil and gas assets in the seller's IM are realistic. This can be achieved quickly and effectively using one of the many black oil correlation applications that are online or available for download. An example of such an "app" is shown in Fig. 9.11, which provides numerous routines to quickly calculate approximate PVT values for oil, gas, and water properties at different surface and reservoir pressures and temperatures.

Among the stated PVT properties to be confirmed, are the gas expansion factor and the oil formation volume factor, which are inputs to the PIIP estimates. The latter parameter is strongly dependent on the GOR, which can be confirmed independently from a check of the produced oil and associated gas volumes. Oil and water viscosities, which influence the fluid mobilities, can be checked to confirm claimed water flood displacement efficiency. Buyers should check bubble point pressure to ensure there is no gas cap present and that reservoir voidage replacement is effective.

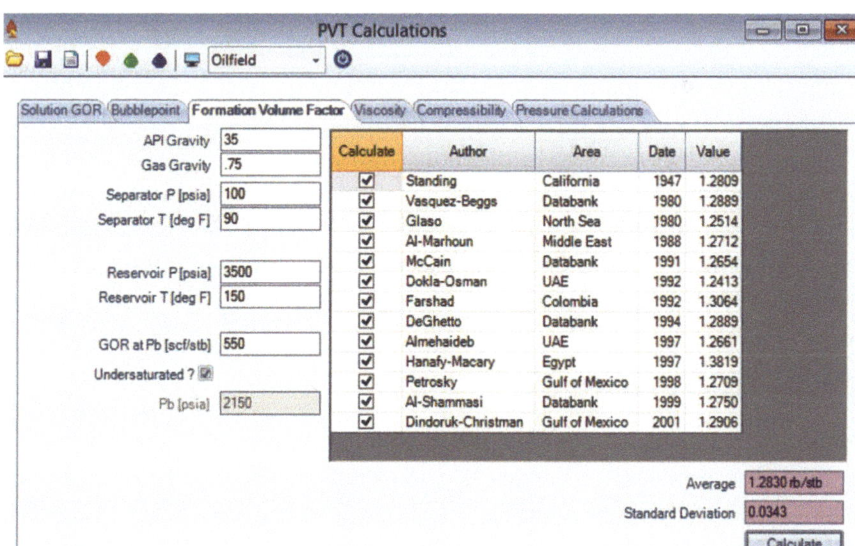


**Figure 9.11** Screenshot of PVT checker provided in the "RE Toolkit" app. *(By permission of Petroleum Solutions Ltd.)*

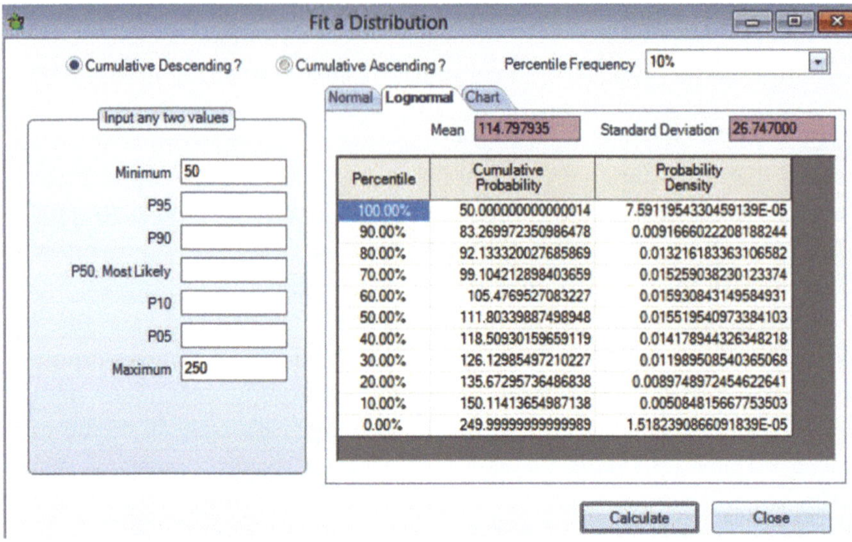


**Figure 9.12** Screenshot of distribution fitter provided in RE Toolkit app. *(By permission of Petroleum Solutions Ltd.)*

## 9.7 Probability distribution fitter

It is essential due diligence that the seller's ranges and distributions of parameters that are input to the PIIP calculations are confirmed. This is readily done by programs or apps that are online or available for download. An example of such an application to fit a probability distribution is shown in Fig. 9.12, where a full probability distribution for the parameter of interest, for either normal or log normal character, can be output by simply inputting two estimated values only. Thus, buyers can quickly confirm that the inputs to the seller's Monte Carlo simulations and geomodels are realistic and conform to oil and gas industry norms.

## 9.8 Analog checklist

When there is a paucity of data, the oil and gas industry relies on analogs to estimate the potential of yet-to-be discovered or developed resources and assumes that the chosen reservoir is analogous to the asset being evaluated, and can provide estimates of reservoir parameters, well performance, and recovery expectations. An ideal analog is a developed reservoir with comparable fluid and rock and fluid properties, reservoir conditions, and drive mechanisms. Analogs should have well-documented physical parameters and adequate performance histories to aid prediction of the qualities of a nearby undeveloped reservoir in the same formation, and assuming the same FDP and operating scenario will be followed (Hodgin and Harrell, 2006). Simply claiming the properties of a field are analogous because of its close proximity to the field of interest is not recommended.

However, selecting an analog is not just limited to similarities in subsurface properties, the facilities requirements, well completion and spacing, artificial lift, and reservoir voidage replacement all help to confirm whether the choice is appropriate, especially when using the chosen analog to forecast the production profile over time, rather than EUR.

The reported increased dependence on analogs to forecast reserves, especially in the United States, is due to the ever-increasing development of unconventional plays in the country. However, it is interesting to note that Lee (2018), a proclaimed expert in such matters, states that analogy may not work in unconventional reservoirs, as poor wells offset good wells, and that no shale resource is an analog for any other shale. Despite these misgivings and the subjectivity of the analogs themselves, the use of analogs is permitted with proven undeveloped (PUD) reserves and their associated higher levels of risk.

Buyers should insist that any PUD reserves claimed by the seller must be supported by best use of all available subsurface and production data, sound engineering judgment, and the selection and application of analogous reservoirs to estimate recovery efficiencies. Yet, how does a potential buyer know if the seller's selected analog is appropriate?

## 9.8.1 Is the seller's chosen analog appropriate?

It is generally accepted that one should not select the best-performing reservoir in a particular region as the basis of comparison to the subject reservoir under study. Yet some sellers insist on doing so as they wish to present their assets on offer in the most favorable light.

When selecting an analog, one should always question whether one really understands the variations in performance in surrounding fields as they relate to the subject reservoir under evaluation. The suitability of a reservoir to serve as an analog is directly related to the purpose of the comparison; hence, one must establish the intended use of the analog. The deliverable required may be EUR per well; minimum well decline, abandonment pressure, drainage radius or well spacing, or expected drive mechanism. Admittedly, analogs are most often used to assign a recovery factor for a given drive mechanism to an estimated PIIP. However, even recovery efficiency can be expressed in different ways, such as per well, per acre-foot at a field level, or even as a function of cumulative injected volume.

Some resource reporting systems offer limited guidance on choosing analogs, merely stating that their selection is defined by features and characteristics of the reservoir, and then provide an exhaustive list of subsurface parameters, without describing how to decide which are suitable. This subjectivity is not shared by other more prescriptive resource reporting systems which insist that, for each PUD well location, the seller should be able to demonstrate beyond reasonable doubt that the claimed drainage areas are "technically justified." The PUD locations must be above the lowest known hydrocarbon or structurally above an observed or estimated hydrocarbon water contact, which has been determined with documented "reliable technology." The seller should also have all the required permits, internal and external approvals, and financing arranged to implement the project within five years. And it does not end there, for every PUD location some systems ask the seller to demonstrate proof of reservoir continuity and economic producibility.

Buyers would do well to heed the more prescriptive advice on selecting analogs when performing their due diligence, whichever resource reporting system they are using.

### 9.8.2 Has the seller demonstrated reservoir continuity?

For a chosen analog to be suitable, the buyer should check that there are good correlations of the productive intervals within the reservoir, from the proven producing areas to the claimed PUD areas. Also, there should be seismic continuity of the productive intervals, which should tie in with the top of the effective porosity as seen by well logs. However, if the seismic interpretation is tied to a marker instead of the producing interval, or the seismic data are discontinuous, or fails to coincide with the reservoir quality in the field itself or in analog fields, then this would increase uncertainty in the estimated reservoir volume significantly.

The seller's assumed depositional model should be consistent with the mapped reservoir parameters and not display highly variable reservoir properties. Pressures recorded across the reservoir can confirm hydraulic continuity if partial depletion is observed in PUD areas, but they may also reveal potentially sealing faults and compartmentalization.

### 9.8.3 Has the seller demonstrated economic producibility?

As described in Appendix 4, the SEC concerns itself with economic producibility, where "economic" simply means a project makes $1 profit more than what was spent, whereas other resource reporting systems are more concerned with project commerciality at the company's corporate rate of return.

Whatever the focus, for analogy to be confirmed, all resource reporting systems require the seller to demonstrate with "reasonable certainty" that the claimed PUD locations are analogous to producing wells, based on data that controls reservoir producibility, such as permeability, net thickness, hydrocarbon saturation and type, and effective porosity.

## 9.9 Analog selection tools

Selecting analogs is a highly subjective process and is often a bone of contention between sellers and buyers as it can have a major impact on the price to do the deal. Hence, it is useful to have tools that can transform words, which can have different meanings to different parties, into a visual representation or a characterizing number that better describes the suitability of an analog. To this end, a couple of techniques are proposed: the radar plot and the comfort factor (CF).

The radar plots shown in Fig. 9.13 allow a graphic comparison of possible analog plays to the target play. In this example, the key subsurface characteristics of two prolific shale gas plays in the United States, the Marcellus and the Eagle Ford, are plotted on top of the parameters of the target unconventional play, the under appraised Bowland Shale in the United Kingdom. While admittedly not definitive, clearly the radial mapping of the Marcellus properties appears to be a better match to those of the Bowland Shale than the Eagle Ford.

The radar plot is relatively straightforward to do in a spreadsheet and can be used as the first stage of a screening process to select a suitable analog. The next stage would be to look at each of the key reservoir characteristics in turn for the remaining shale plays, which can be done using CFs.

The CF technique computes the area of overlap for the analog and the target triangular probability distributions, which define the minimum, most

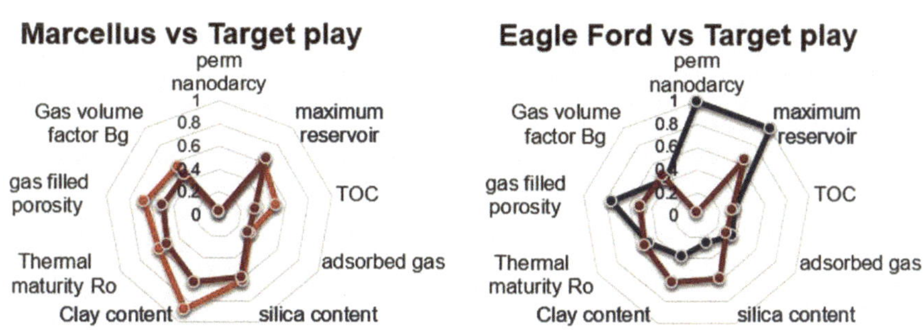


**Figure 9.13** Comparing radar plots of shale play parameters to select potential analog for the target shale play.

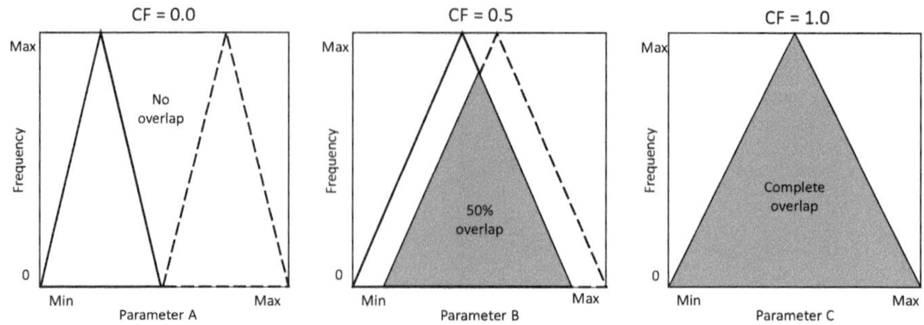


**Figure 9.14** Comparing overlapping areas of triangular probability distributions of play parameters to derive comfort factors that can aid selection of a potential analog for the target play. *(After Alvarado et al. (2008).)*

likely, and maximum values for each parameter of interest. Fig. 9.14 shows the range of no analogy (CF of zero) to full analogy (CF of one).

The resulting CF values for each of the parameters are combined to give an overall CF for each of the potential analogs. Those candidate plays with the highest overall CF values are considered for selection as suitable analogs for the target play.

## 9.10 Back to the future with DCA

The industry looks back at the past production history of wells and fields to predict their future production and so estimate their recoverable volumes. Decline curve analysis (DCA) was formally introduced to the oil and gas industry by J.J. Arps, a geologist-cum-engineer, in the 1960s, and it remains an important instrument in the reserves auditor's toolkit.

Fundamentally, DCA uses a mathematical relationship that describes the rate at which oil or gas production from a well or a field decays over time. Once a well exceeds its transient flow period, when the pressure of the reservoir region being drained is virtually constant, it will then "see" an actual reservoir flow boundary, and the reservoir pressure will start to decline. This is the time interval when DCA can be employed. The time it takes for a producing well to "feel" the limits of its drainage area can be a few minutes to several years, depending upon permeability and the areal extent of the conventional reservoir.

The results of DCA can be supported by ratio analyses of past produced phase volumes, such as the observed trends in water oil ratio, gas oil ratio, liquid gas ratio, water cut, and oil cut, which are plotted against cumulative petroleum production.

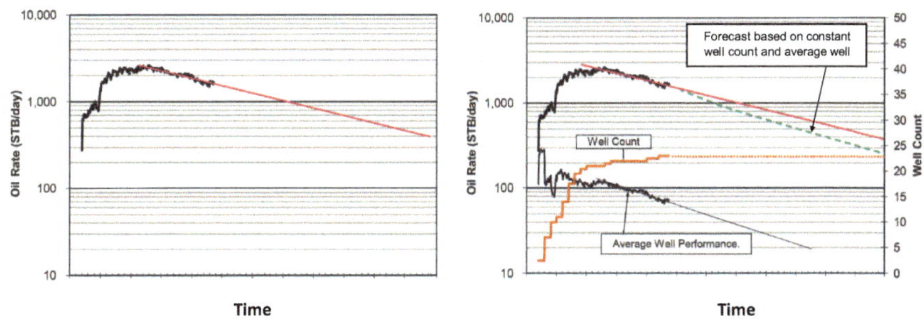


**Figure 9.15** Using average well performance mitigates against infill-drilling programs leading to optimistic recovery estimates from decline curve analysis at an aggregated field level. *(After Harrell et al. (2004).)*

This final section considers the use and abuse of these traditional production history matching techniques and the impact of failing to take past experience into account when making forecasts of the future performance and recovery from wells and fields.

## 9.10.1 Composite field DCA is often optimistic

It is always preferable to perform DCA at an individual completion interval or well level, but very often a seller will provide an asset's production history in an aggregated form, such as at a lease, reservoir unit, or field level.

An example of a field production history is depicted in Fig. 9.15 (*left*), which does show a continual decline over time, which appears to be well defined and so provides some confidence that a mathematical "best fit" would allow future field performance to be predicted.

However, performing DCA at field level to forecast EUR is only valid if the field's well count is stable, the producing well operating conditions remain largely unchanged, and any planned well remedial work can be regarded as routine maintenance. If one plots the field's well count against rate history, as shown in Fig. 9.15 (*right*), it is clear that the predicted profile will be unattainable without an ongoing drilling program to supply extra wells at the same frequency and with similar results as those used in the DCA. Also, by continually bringing new wells on line, the true decline in field performance will be masked.

If the seller cannot provide individual well production histories for DCA, the buyer can try to offset the masking effect of the continual drilling program in the field by restating well data as an average monthly production per well (Harrell et al., 2004). Performing DCA on these averaged data should provide a more reliable forecast for the field.

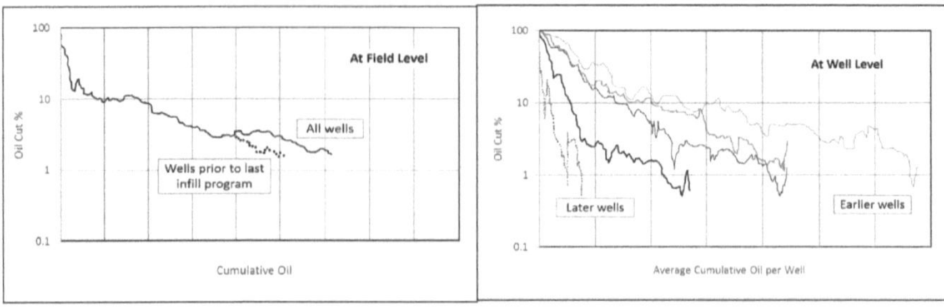


**Figure 9.16** Using oil cut trends to identify interference by analogy, at field level and at well level. *(After EPIC.)*

### 9.10.2 Using oil cut to identify interference between producing wells

Buyers must exercise caution when reviewing a seller's estimates of recoverable volumes which take into account new infill wells in the field. If the seller's assumed well spacing is too small, then there may be interference between producers, and the estimated incremental recovery from the new wells may not be realized. As shown in Fig. 9.16, it is possible to use the decline trends in oil cut, which is simply 100% minus the water cut, in a qualitative way to identify interference between producing areas, both at a field level and at a well level.

With regard to analysis at field, the buyer can compare the performance of different groups of wells that are classified by "vintage" or by the year that they were brought on stream as depicted in Fig. 9.16 (*left*). The trend in oil cut for the wells that were present before the last infill program is plotted alongside the same trend for all wells and the former oil cut trend can be checked for any steepening decline that would suggest interference.

At the well level, one can also plot the average oil cut trends of different vintage well groups in the field versus individual well cumulative oil production, as shown in Fig. 9.16 (*right*), and look for a steepening of slope in the earlier vintage well groups. If interference between producing areas is indicated, the buyer should consider reducing the seller's claimed EUR values.

## 9.11 Thiessen polygon mapping

Thiessen polygon maps, which are also called Voronoi diagrams, are used to define and to delineate proximal regions around individual data points by using polygonal boundaries. They are employed by many disciplines including weather forecasters, hydrologists and geographers, and by

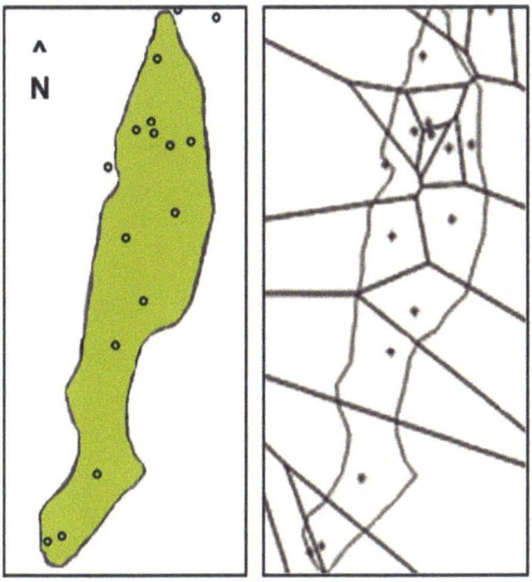


- Compute areas of 14 polygons
- Apply PP net pay for each well
- Sum polygon net pays = PIIP

| | Petrel / Polygon |
|---|---|
| Scalar | 0.947 |

| Area | Well | Area km2 | Polygon Area contrib | Main STOIIP MMstb | Polygon STOIIP contrib | Main Scaled MMstb |
|---|---|---|---|---|---|---|
| northern | 1 | 1.90 | 1.2% | 6.8 | 1.6% | 6.4 |
| northern | 2 | 9.75 | 6.3% | 56.6 | 13.8% | 53.7 |
| northern | 3 | 5.87 | 3.8% | 59.0 | 14.3% | 55.8 |
| northern | 4 | 8.39 | 5.5% | 55.3 | 13.4% | 52.4 |
| northern | 5 | 11.69 | 7.6% | 110.1 | 26.8% | 104.3 |
| northern | 6 | 7.39 | 4.8% | 24.8 | 6.0% | 23.5 |
| northern | 7 | 5.58 | 3.6% | 12.6 | 3.0% | 11.9 |
| southern | 8 | 16.03 | 10.4% | 21.0 | 5.1% | 19.9 |
| southern | 9 | 18.59 | 12.1% | 1.9 | 0.5% | 1.8 |
| southern | 10 | 16.52 | 10.7% | 11.2 | 2.7% | 10.7 |
| southern | 11 | 20.66 | 13.4% | 47.6 | 11.6% | 45.1 |
| southern | 12 | 22.51 | 14.6% | - | 0.0% | - |
| southern | 13 | 7.64 | 5.0% | 4.4 | 1.1% | 4.2 |
| southern | 14 | 1.28 | 0.8% | 0.5 | 0.1% | 0.4 |

**Figure 9.17** Workflow shows going from an oil field top structure map with well locations, to a corresponding Thiessen polygon map, then to a table which provides a PIIP estimate within 5% of the geomodel PIIP.

geologists in the mining sector, who use them to estimate resource volumes from exploratory boreholes.

Polygon maps can be used in the data room to create relatively accurate PIIP estimates from the seller's base geomodel. The subsequent model, which is based on reservoir area around each of the wells and extending out to the field oil water contact (OWC), can be used in numerous ways for rapid sensitivity analysis and "what-if" scenarios.

Fig. 9.17 shows a polygon map and outcome for an oil field after the individual well petrophysical analysis results have been applied. The polygon mapping estimate was within 5 percent of the geomodel estimate. The size of polygons in the table gives an idea of how much of the in-place volume is influenced by each well. On an areal basis one might think the southern polygons are more important, but after having applied the relevant petrophysical results it is obvious that the northern polygons contain the lion's share of the PIIP, and polygon-5 in particular.

Note that any major structural faults can be used as boundaries for the polygons (i.e., well parameters of a polygon on one side of the fault are not applied to the polygon on the other side of the fault). License boundaries can also act as boundaries for the polygons, although the polygon crossing the lease line would have the same parameters on either side. This allows PIIP to be broken down by lease and so the resultant polygon map can be used for computing working interest based on PIIP in unitization studies.

The Thiessen polygon model permits different fluid properties to be assigned to different polygons if hydrocarbon quality varies across structure. Different well performance (and associated EUR) can be allocated to different polygons to reflect spatial variation in reservoir quality. Also, the field area itself can be altered if the OWC is uncertain by increasing the areas of the outer polygons accordingly.

The main advantages of a polygon mapping approach (which has been "adjusted" to tolerably match the geomodel's base PIIP) are flexibility and speed to allow the M&A team to make rapid and reasonable judgments on the relative importance of various assumptions made by the seller.

## 9.12 Bubble maps to review reservoir drainage and sweep

Buyers can plot individual well production volumes, both current and forecast from DCA and ratio analysis, as circles, rectangles, or ellipses depending on the well geometry. The dimensions of the appropriate shape are set by the petrophysical parameters of the well and by the produced fluids' formation volume factors. The technique has proved extremely useful in identifying or confirming infill-drilling opportunities.

If one considers the case of a vertical producing well in an oil reservoir, the equation shown below is the STOIIP calculation solved for an equivalent drainage radius ($r_d$, in ft) for the produced oil and water phases.

$$r_d = \sqrt{\frac{(O_{cum} \times B_o + W_{cum} \times B_w) \times 5.615}{\pi h \phi (1 - S_{wi} - S_{or})}}$$

In the rearranged equation above, $O_{cum}$ and $W_{cum}$ are the cumulative production of oil and water (in barrels) and $B_o$ and $B_w$ are the formation volume factors of oil and water ($rb/stb$), respectively, $h$ is the net pay (ft), $\emptyset$ is the effective porosity (fraction), $S_{wi}$ denotes the irreducible water saturation (fraction), and $S_{or}$ is the residual oil saturation (fraction). An example of a bubble map for a mature gas field is shown in Fig. 9.18. Unlike the oil bubble plot which works on movable oil volume, for a gas field, the bubble size corresponding to each individual wells' gas production is scaled by the gas expansion factor at the assumed abandonment pressure.

If the producing well's orientation is highly deviated or horizontal, its drainage area can be approximated with a rectangle as shown in Fig. 9.19.

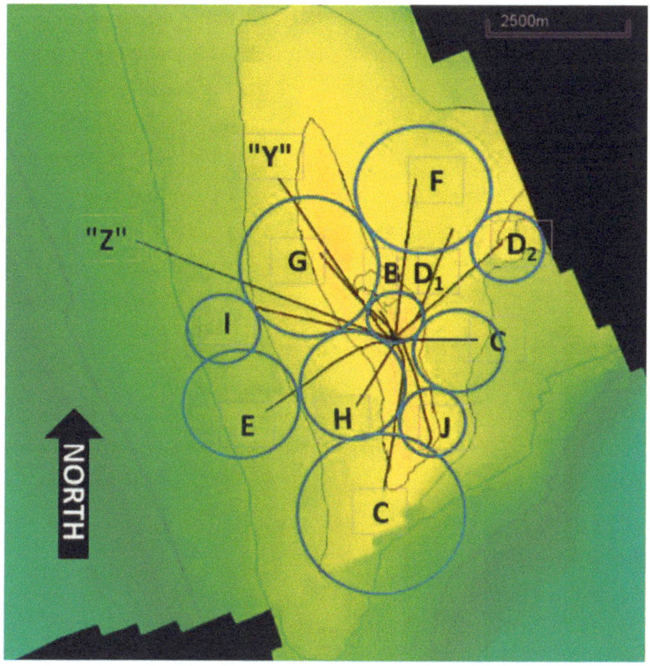


**Figure 9.18** Bubble map of circular drainage areas based on EUR for vertical wells in a mature gas field. Of the two infill targets, "Y" is just off the crest by maybe partially depleted by producer G, whereas and "Z" on the western flank may be OK, but closer to the gas water contact.

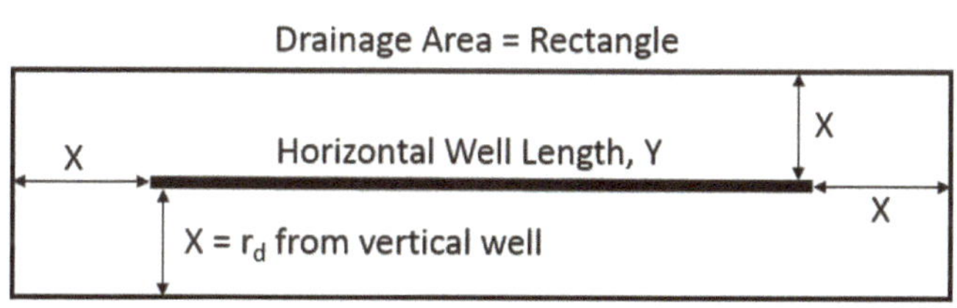


**Figure 9.19** Rectangle is an appropriate shape for drainage area of a horizontal well.

The distance from the horizontal drain hole to each side of the rectangle is often taken to be the drainage radius estimated from a vertically oriented well.

## 9.13 Buyer's own simpler versions of seller's models

It is often the case that sellers do not provide working versions of their geomodels or dynamic simulation models to potential buyers. The seller's models can be inspected in the PDR and their respective inputs and results are usually loaded on to the workstation for review in the VDR. However, it is generally the case that buyers fail to get possession of the seller's working models, which would allow them to run their own sensitivities.

To mitigate against this, buyers may consider building their own cruder versions of the static and dynamic models that approximate the spatial of 3D variations in rock and fluid properties. Time is of the essence in a due diligence exercise, but these simpler models are relatively quick to create. The seller's own model input data are incorporated and then adjusted to match the seller's base case results. These simpler representations of the seller's assets can then be run many times, very quickly, to see what might happen if certain key parameters or assumptions were to change or be questioned. The difference between the result of a sensitivity run and the reference case (which mimics the seller's base case) should provide a tolerably accurate view of what would happen to the base case result if a particular parameter was changed. This knowledge would allow a potential buyer to organize his opening negotiating stance and to formulate his bidding strategy.

If we consider a seller's discovery undergoing appraisal which has its own geomodel, a potential buyer could use his interrogation of the latter (in the PDR) to develop a Monte Carlo simulation equivalent. This much simpler probabilistic model should have its P50 or most likely result adjusted to match that of the geomodel base case via manipulation of its input parameter ranges and distributions. Similarly, the buyer's Monte Carlo model P90 and P10 results can be tweaked to agree with the low and high cases of the seller. An example of Monte Carlo simulation output is shown in Fig. 9.20. The software shown allows input of rock and fluid parameter ranges and distributions, top and base structural surfaces, fluid contact depths, and saturation height functions, which permits significant geomodel uncertainty to be investigated quickly and reasonably accurately using this simpler approach.

In a similar fashion, a producing asset can be represented by simple one-dimensional model that can be history matched and then adjusted to mimic the results of the seller's field's base EUR. Fig. 9.21 illustrates how such software can model the water displacement of oil in a 10-layer representation of the field, complete with production history match prior to running a forecast. The 1D model inputs can then be altered to reflect the various uncertainties and concerns held by the potential buyer. The difference in performance from each sensitivity run and the reference case of the 1D model should provide a reasonable estimate of how the seller's FFM would behave if it underwent similar parameter changes.

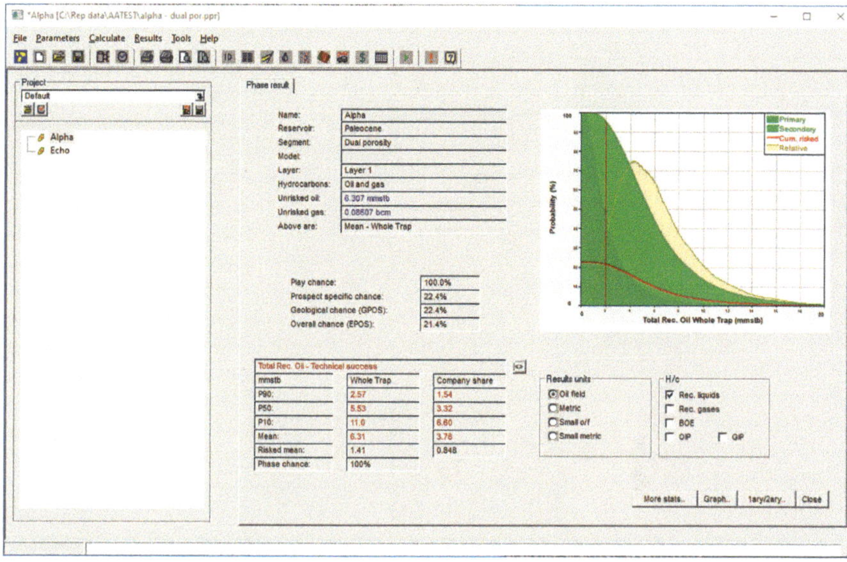

**Figure 9.20** Screenshot of REP software for building a "scaled" Monte Carlo simulation of the seller's geomodel results and subsequently used to study subsurface sensitivities (Logicom, 2018). *(By permission of Logicom E&P Ltd.)*

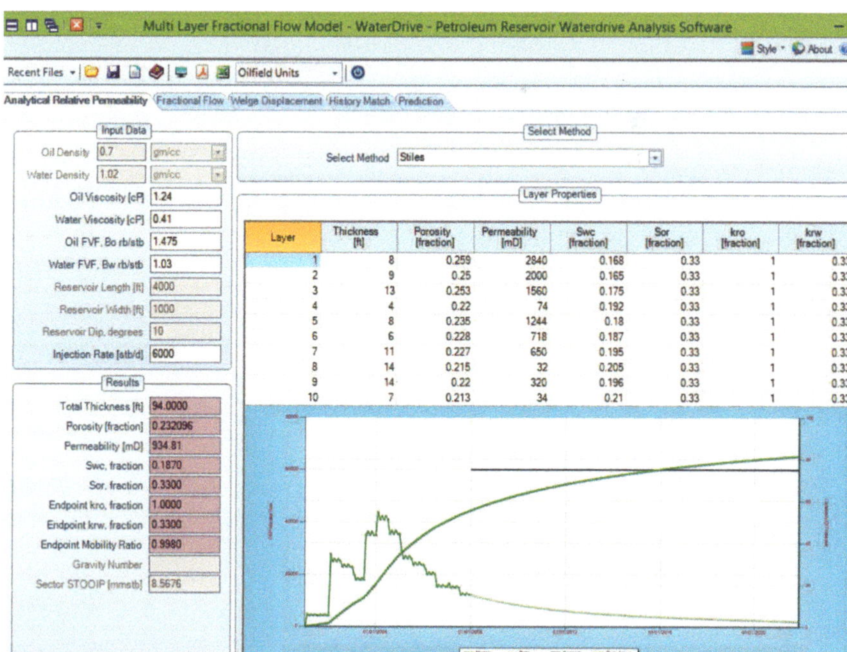

**Figure 9.21** Screenshot of WATERDRIVE application used to build a history matched multilayer 1D displacement model of the seller's FFM, which was then used for "what-if" scenarios (Petroleum Solutions, 2018b). *(By permission of Petroleum Solutions Ltd.)*

# Glossary

| | |
|---|---|
| **1P** | Proven reserves = P1, deterministic scenario estimate |
| **2P** | Proven + Probable reserves = P1 + P2, deterministic scenario estimate |
| **3P** | Proven + Probable+Possible reserves = P1 + P2 + P3, deterministic scenario estimate |
| **AAPG** | American Association of Petroleum Geologists |
| **ABEX** | Abandonment expenditure; cost of decommissioning |
| **ACQ** | Annual Contract Quota; in a GSA |
| **AFE** | Authorization for Expenditure |
| **API** | American Petroleum Institute |
| **BCGA** | Basin-centered gas accumulation |
| **BDP** | Best demonstrated practice |
| **Bg** | Gas formation volume factor (rb/scf) |
| **Bo** | Oil formation volume factor (rb/stb) |
| **BOD** | Basis of design |
| **BOE** | Barrels of oil equivalent; for gas volumes in oil heat capacity terms |
| **BOP** | Blowout preventer |
| **BS&W** | Bottom sand and water |
| **Btu** | British thermal unit |
| **C&DI** | Compliance and Disclosure Interpretation |
| **C&I** | Cost and insurance |
| **CA** | Confidentiality agreement |
| **CAPEX** | Capital expenditure; cost of development drilling and infrastructure |
| **CBM** | Coal bed methane |
| **CEC** | Cation exchange capacity; used to compute Sw in shaly sand |
| **CF** | Comfort factor; used when selecting analogs |
| **Cfge** | Cubic feet of gas equivalent; for oil volumes in gas heat capacity terms |
| **CGR** | Condensate gas ratio (stb/MMscf) |
| **CHOPS** | Cold heavy oil production with sand |
| **COGEH** | Canadian Oil & Gas Evaluation Handbook |
| **COP** | Cessation of production, when a field reaches its economic limit |
| **CPI** | Computer processed interpretation; usually of log data |
| **CPR** | Competent Person's Report |
| **CSA** | Canadian Securities Agency; Canada regulator |
| **CV** | Calorific value, heating capacity of natural gas |
| **D** | Dry hole cost |
| **D&C** | Drilling and completions |
| **DAS** | Distributed acoustic sensing; in well fiber optic monitoring |
| **DCA** | Decline curve analysis |
| **DCF** | Discounted cash flow, future revenues and costs discounted to today |
| **DCQ** | Daily Contract Quota; in a GSA |
| **DCS** | Distributed chemical sensing; in well fiber optic monitoring |

| | |
|---|---|
| **DEVEX** | Development expenditure, including drillex and capex |
| **DF** | Fluid density (gm/cc), used to calculate porosity from density log |
| **DG** | Decision gate |
| **DL** | Density log reading (gm/cc) |
| **DM** | Matrix density (gm/cc), used to calculate porosity from density log |
| **DoC** | Declaration of Commerciality |
| **DPIR** | Discounted profit to investment ratio |
| **DPUQ** | Drilling production utilities quarters; offshore platform configuration |
| **DRILLEX** | Drilling expenditure; costs directly related to drilling operations |
| **DST** | Drill stem test |
| **DTS** | Distributed thermal sensing; in well fiber optic monitoring |
| **E** | Gas expansion factor (scf/rcf) |
| **EIA** | Energy Intelligence Administration (USA) |
| **EMV10** | Expected monetary value at 10% discount rate |
| **EOR** | Enhanced oil recovery |
| **EOS** | Equation of state; to model PVT behavior of petroleum fluids |
| **EPIC** | Engineering, procurement, installation, and construction; common form of contracting in offshore sector |
| **EUR** | Estimated ultimate recovery |
| **FBHP** | Flowing bottom hole pressure |
| **FBHT** | Flowing bottom hole temperature |
| **FID** | Final investment decision |
| **FDP** | Field development plan |
| **FFM** | Full field model; for dynamic reservoir simulation |
| **FWHP** | Flowing well head pressure |
| **FWHT** | Flowing well head temperature |
| **FWL** | Free water level; depth in reservoir below which Sw is 100% |
| **GIIP** | Gas initially in place |
| **GOC** | Gas oil contact |
| **GOM** | Gulf of Mexico |
| **GOR** | Gas–oil ratio (scf/stb) |
| **GR** | Gamma ray; logging tool response to natural radioactivity in formation |
| **GRV** | Gross rock volume (of reservoir) |
| **GSA** | Gas Sales Agreement |
| **HKW** | Highest known water |
| **HPHT** | High pressure high temperature |
| **HPV** | Hydrocarbon pore volume |
| **HSE** | Health, safety, and environment |
| **HWC** | Hydrocarbon water contact; depth in reservoir below which only water can move |
| **IM** | Information memorandum |
| **Kh** | Horizontal permeability |
| **Kr** | Relative permeability |
| **Kv** | Vertical permeability |
| **LAS** | Log ASCII Standard; log data format |

| | |
|---|---|
| **LHS** | Latin hypercube sampling |
| **LKH** | Lowest known hydrocarbon |
| **LNG** | Liquefied natural gas |
| **$m^3$** | cubic meters |
| **M&A** | Mergers and acquisitions |
| **MM** | Million |
| **MT** | Measured thickness |
| **NCS** | Norwegian continental shelf |
| **NDA** | Non-disclosure agreement |
| **NGL** | Natural gas liquid |
| **NPD** | Norwegian Petroleum Directorate; Norway's regulator |
| **NPI** | Net profit interest |
| **NPorV** | Net pore volume of reservoir |
| **NPV10** | Net present value at 10% discount rate |
| **NRV** | Net rock volume of reservoir |
| **NTG** | Net to gross |
| **OBM** | Oil-based mud |
| **OCM** | Operating Committee Meeting |
| **ODT** | Oil down to |
| **OGA** | Oil and Gas Authority; UK's regulator |
| **OPEX** | Operating expenditure; cost of operations |
| **ORB** | OPEC Reference Basket; oil price marker crude |
| **ORRI** | Over-riding Royalty Interest |
| **OWC** | Oil water contact |
| **P1** | Proven reserves = 1P, deterministic estimate from incremental method |
| **P2** | Probable reserves ≠ 2P, deterministic estimate from incremental method |
| **P3** | Proven reserves ≠ 3P, deterministic estimate from incremental method |
| **P50** | Median, also called best estimate, can be exceeded with 50% probability |
| **P&ID** | Piping and instrumentation diagram |
| **Pbp** | Bubble point pressure |
| **Pc** | Capillary pressure |
| **Pd** | Chance of development |
| **PDC** | Polycrystalline diamond cutter; type of drill bit |
| **PDR** | Physical data room |
| **PE** | Production efficiency |
| **PFD** | Process flow diagram |
| **Pg** | Chance of geologic success |
| **PHI** | Porosity; reservoir pore space that can hold fluids |
| **Pi** | Initial reservoir pressure |
| **PIIP** | Petroleum initially in place |
| **PLT** | Production logging tool; measures flow profiles in producing wells |
| **Pmin** | Minimum exploration risk |
| **PRMS** | Petroleum Resource Management System |

| | |
|---|---|
| **PSC** | Production Sharing Contract |
| **psi** | Pounds per square inch; measure of pressure |
| **PUD** | Proven undeveloped |
| **PVT** | Pressure volume temperature, to describe petroleum fluid behavior |
| **QC** | Quality control |
| **QPR** | Qualified persons report |
| **Rb** | Barrels of oil at reservoir conditions |
| **RCA** | Routine core analysis |
| **Rcf** | Reservoir cubic feet |
| **RF** | Recovery factor |
| **Rh** | Horizontal displacement efficiency |
| **Rm** | Microscopic displacement efficiency |
| **Rsi** | Initial solution gas ratio |
| **Rv** | Vertical displacement efficiency |
| **RVP** | Reid vapor pressure |
| **Rw** | Formation water resistivity (ohm m), used to calculate Sw |
| **SCAL** | Special core analysis |
| **Scf** | Standard cubic feet; at defined surface conditions of pressure and temperature |
| **SEC** | Securities & Exchange Commission; US regulator |
| **SF** | Shrinkage Factor; of wet gas as it is produced to surface |
| **SHF** | Saturation height function |
| **Soi** | Initial oil saturation |
| **Sor** | Residual oil saturation |
| **SPE** | Society of Petroleum Engineers, authors of PRMS |
| **SPEE** | Society of Petroleum Evaluation Engineers; authors of COGEH |
| **Stb** | Stock tank barrels of oil; at defined surface conditions of pressure and temperature |
| **STEO** | Short-Term Energy Outlook (from EIA) |
| **STOIIP** | Stock tank oil initially in place |
| **Sw** | Water saturation; in reservoir pore space |
| **TCM** | Technical Committee Meeting |
| **TOC** | Total organic content; of source rock |
| **TPR** | Time pressure rate; well test gauge data format |
| **Tres** | Reservoir temperature |
| **TRV** | Technically recoverable volume; no commercial constraints applied |
| **Ts** | Surface temperature |
| **TVD** | True vertical depth; datum for subsurface studies |
| **TVDSS** | True vertical depth sub sea; datum for subsurface studies |
| **TVP** | True vapor pressure |
| **TVT** | True vertical thickness |
| **UFD** | Utility flow diagram |
| **UKCS** | United Kingdom Continental Shelf |
| **V** | Permeability variation coefficient; from Dykstra–Parsons analysis |
| **VDR** | Virtual data room |
| **VSP** | Vertical seismic profile |
| **WACC** | Weighted average cost of capital; used as corporate hurdle rate |

| | |
|---|---|
| **WGIIP** | Wet gas initially in place; used in gas condensate fields |
| **WOR** | Water oil ratio (stb/stb) |
| **WP&B** | Work program and budget |
| **WPC** | World Petroleum Congress |
| **WS** | Waxman–Smits; method to calculate Sw |
| **WTI** | West Texas Intermediate; US oil price, marker crude |
| **WUT** | Water up to |

# Appendix 1: QC of inputs to estimates of PIIP and TRV

Every M&A venture will require due diligence to be performed on one or several estimates of petroleum initially in place (PIIP) and the associated technically recoverable volumes (TRVs) of petroleum. This aim of this appendix is not only to offer some background on the corresponding equations used to evaluate in place and recoverable volumes of conventional reservoirs but also to provide typical ranges and distributions for the input parameters used in the evaluations. The seller can use this information to ensure his input data for the assets on offer look reasonable to potential acquirers. Buyers can check the data ranges and distributions provided in the data room and ask the seller to justify any that fall outside the accepted norms.

The different reservoir drive mechanisms and how they influence recovery, production character, and field development are discussed. These should allow buyers to quickly confirm whether the input parameters used in a seller's estimates are realistic.

Let us begin by considering the generic volumetric formulae for estimating PIIP, which are quite straightforward for conventional reservoirs, with the in situ resource volumes being reported at surface conditions.

## A1.1 Estimating in place conventional petroleum resource volumes

### A1.1.1 Stock tank oil initially in place

For a conventional reservoir, the amount of oil in place reported at stock tank conditions at surface is

$STOIIP = C1 \times GRV \times NTG \times PHI \times (1 - Sw) \div Bo$

STOIIP = stock tank oil initially in place (stb)

C1 = units conversion factor for working in barrels = 6.29 stb/m$^3$

GRV = gross rock volume (m$^3$)

NTG = net-to-gross (fraction)

PHI = effective porosity (fraction)

Sw = water saturation (fraction)

Bo = oil formation volume factor (reservoir volume/surface volume)

The formation volume factor accounts for the shrinkage of oil as it is produced at surface. The more gas is dissolved in the oil, the more energy it has within it and the greater it shrinks as it is produced. All of which leads us on to estimating the volumes of associated gas in an oil field.

### A1.1.2 Associated gas

Every oil-bearing reservoir will have some associated gas that is dissolved in the oil; there may even be free gas in an overlying gas cap. To determine the recoverable gas volume associated with an oil discovery, we need to know something about the phase behavior of the oil.

Undersaturated oil has no free gas as it is dissolved in the oil, but as reservoir pressure is depleted to below its bubble point pressure, the dissolved gas is released from the reservoir crude. When producing this type of oil, operators strive to maintain the reservoir pressure above the oil's bubble point by injecting water or gas or both. Hence, the amount of solution gas produced along with the oil, which is measured using the gas to oil ratio (GOR), should remain constant, so that

Associated gas in place (scf) = STOIIP (stb) × GOR (scf/stb)

Associated gas recovery for full voidage replacement waterflood (scf) = TRV of oil (stb) × GOR (scf/stb)

Saturated oil is already at its bubble point pressure in the reservoir, so this crude will release gas immediately if there is a pressure drop in the reservoir. However, when the oil is saturated and the reservoir pressure is allowed to fall below its bubble point, the producing GOR will vary with time. The determination of the associated gas recovery is far more difficult, but it should not be ignored in a due diligence exercise.

### A1.1.3 Gas initially in place

The volumetric formulae for estimating the amount of gas in a conventional reservoir, reported at standard conditions at surface, is

$$GIIP = C2 \times GRV \times NTG \times PHI \times (1 - Sw) \times E$$

GIIP = gas initially in place (Bscf)

C2 = units conversion factor for working in cubic feet = $35.314 \times 10^{-9}$ (Bscf/$m^3$)

E = gas expansion factor (surface volume/reservoir volume)

A barrel of reservoir oil shrinks when it is produced to surface as the associated gas dissolved in it is released, but a volume of reservoir gas expands when it is produced to surface as its confining pressure decreases. Hence, the formation volume factor for gas is an expansion term in the numerator of the GIIP equation.

### A1.1.4 Associated condensate

Although gas expands when it is produced to surface, there is also a small shrinkage term due to the chemical composition of the gas changing due to decreasing pressure and temperature in the production tubing in the well, which results in the appearance of a liquid phase known as condensate. Thus, a shrinkage factor (SF) must be applied to the produced volume of raw or wet gas to give an equivalent produced volume of dry gas plus condensate. Some gas fields contain very dry gas with negligible vaporized liquid, which is usually reported as a low condensate gas ratio (CGR) that is often reported in stb of condensate per MMscf of gas. With low CGR values, in single figures, SF is close to unity and is usually ignored in GIIP estimates.

However, one should not ignore condensate volumes in very wet gas, known as gas condensate, as the CGR values are much higher. The cut-off between dry gas and wet gas is traditionally taken to be around a reported CGR value of 30 stb/MMscf, which corresponds to a raw gas volume shrinkage of around 2 percent.

Unlike the associated gas with an oil field, the associated condensate cannot be estimated by multiplying the GIIP estimate by the initial producing CGR value. As a gas field is produced, its reservoir pressure depletes and the gas composition is continually changing. Hence, the producing CGR is also continually decreasing with time.

Unless some partial or full gas recycling scheme is envisaged, it is assumed that the gas condensate exhibits retrograde behavior, where liquids drop out of wet gas in the reservoir as pressure drops below the dew point. Thus, the observed initial producing CGR will fall as gas is produced from the reservoir and its pressure is depleted. A rule of thumb is that less than 40 percent of the condensate contained in the wet gas initially in place (WGIIP) is recovered on blowdown, but it can vary quite widely, from 20 percent to 50 percent, depending upon fluid composition, reservoir heterogeneity, and structure. Gas condensate bearing assets need two recoverable volumes to be calculated from the last two steps of the workflow shown in Fig. A1.1.

## A1.2 Estimating PIIP input parameter ranges and distributions and common errors

Ranges for each input parameter of the PIIP calculation are provided by geoscientists in order to make an educated guess of the spread of values within which the true value of in place volume may lie. This range of potential PIIP is wider in assets undergoing exploration and appraisal as they have little or no production history to guide in place estimates.

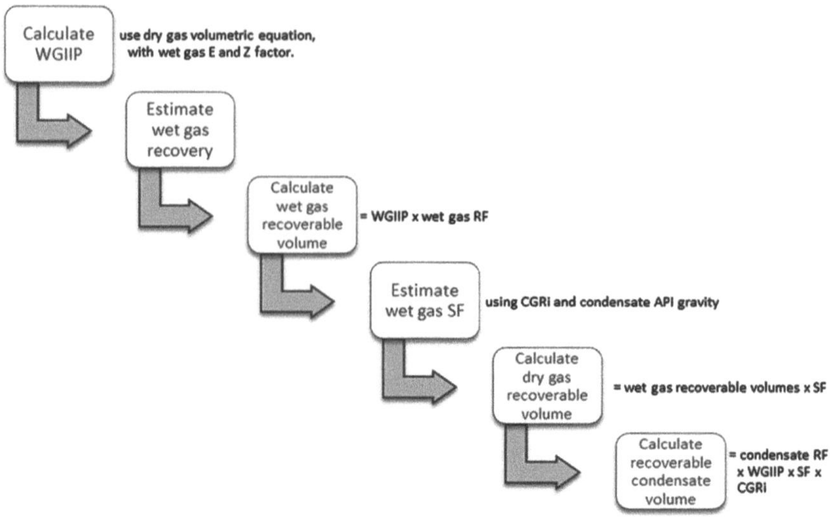


**Figure A1.1** Workflow for estimating gas condensate in place and recoverable volumes.

## A1.2.1 Monte Carlo simulation and choosing parameter distributions

The reservoir parameter ranges and their corresponding distributions are often input to a Monte Carlo simulation, which calculates PIIP thousands of times using a random selection from the input distributions for each variable at each iteration. The results of these numerous scenarios provide a statistical distribution of PIIP from which one can appreciate the associated uncertainty. However, the weakness of Monte Carlo technique lies in the input assumptions, as it is arguable that the answer is guessed before the analysis is undertaken. Therefore, the selection of the appropriate probability distribution for each input variable is paramount, along with deciding which input variables are independent and which ones should be linked with each other to reflect their dependency. Repeatability can deteriorate if the number of simulations is insufficient, so it is recommended that over 30,000 passes are performed to be confident in getting the same result. Most Monte Carlo simulation software offer the Latin Hypercube Sampling (LHS) option, which speeds up the model's convergence and so reduces the

**Figure A1.2** Common input parameter probability distributions used in Monte Carlo simulation.

number of required simulations to around 10,000 passes. Fig. A1.2 shows the most commonly used distributions.

If one is trying to describe variables such as porosity or inflation, it is traditional to use the normal distribution, also known as the bell curve as it is symmetrical about its most likely value, which is also the same as its mean in this case. Users need to define the parameter's mean value and its standard deviation to describe the symmetric variation about the mean.

The log normal distribution should be used for characterizing most subsurface parameters used in volumetrics as they tend to have large variation and be highly uncertain. This distribution allows the representation of values that do not go below zero, but which can have large upside potential. Its use for GRV inputs in exploration prospects is a must. Failure by the seller to do so would constitute a red flag to potential buyers.

It is often tempting to use a triangular distribution, where one simply defines the minimum, most likely, and maximum values. However, there is nothing in nature that is distributed this way, except maybe a witch's hat. It assumes one knows the exact spread of the parameter, which is rarely the case, and it always over predicts the mean when estimating PIIP. Triangular distributions should be avoided.

The uniform distribution, also known as the boxcar distribution, is the simplest of all, where every value of a parameter has an equal chance of occurring, and one only needs to define the range minimum and maximum. Its use suggests that one knows nothing about the variable whatsoever, except its lowest and highest values, yet that may be appropriate when trying to predict what the future oil price might be. Some petrophysicists argue that the use of a uniform distribution might be appropriate when considering where the yet to be penetrated oil-water-contact may be located between an oil-down-to (ODT) depth in one well and a water-up-to (WUT) depth in another well in an underappraised petroleum accumulation. Overall, geoscientists tend to avoid uniform distributions.

Let us now consider each of the input parameters of the PIIP estimate in turn. Rules of thumb for the expected ranges of values for each parameter are given, which can be used for quick checks in the data room. The given ranges are typical, but by no means are they sacrosanct, but if the seller's parameter values fall outside accepted ranges then the buyer should demand to know why. Readers should note that the less one knows about the reservoir under investigation, the larger the parameter range should be.

### A1.2.2 Gross rock volume

By far, the biggest source of error when calculating PIIP is a poorly defined GRV. Lack of subsurface data apart, the characterization of the reservoir will lack the required detail if inappropriate or no cut-offs are used. Failure to identify and acknowledge baffles and barriers, both areally and vertically, will lead to a misplaced assumption that there is vertical continuity throughout the reservoir.

Incorrect reservoir areas are computed whenever subsurface maps are unconstrained by the picked top and base of the effective reservoir.

Fig. A1.3 shows how the upper part of the formation should not be included in the effective reservoir calculation as it fails to meet the applied cut-offs. Similarly, the base of the reservoir is the deepest occurrence of net sand in the well which can be demonstrated to be in the same pressure system. The base of the reservoir defines the lowest known hydrocarbon (LKH), not the base of the formation, so if the base of the formation fails to meet the applied cut-offs, it should be excluded from the effective reservoir (Harrell et al., 2004).

Also, the failure to recognize uncertainties in the structural interpretation and not limiting the proved area will result in overoptimistic reservoir area. Constraints on the areal extent of the reservoir should be applied by strictly imposing fluid contacts along with the recorded depths of the LKH and the highest known water (HKW). If the seller insists on stating larger drainage areas and claiming there is lateral continuity away from existing well control, then as part of due diligence, the M&A team should insist that these assertions can be technically justified.

Erroneous reservoir volumes arise with the incorrect application of true vertical thickness (TVT) when integrating data from deviated wells. The importance of using TVT rather than measured thickness (MT) can make a significant difference when modeling deviated wells in a dipping reservoir. The concept of TVT correction in deviated wells and dipping beds is explained graphically in Fig. A1.4.

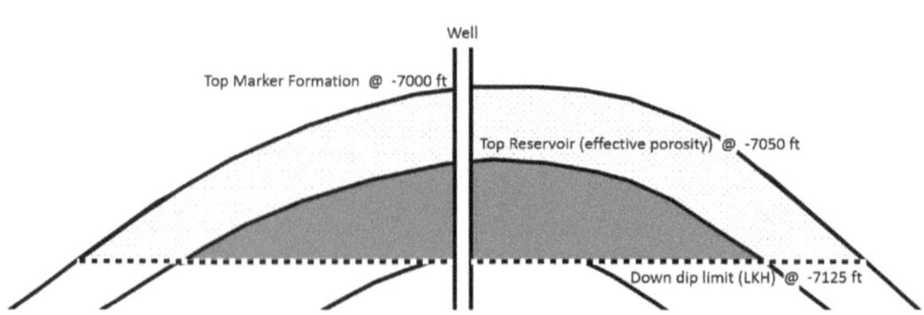


**Figure A1.3** Confusion over whether top and base depths are defining either the formation or the reservoir can cause significant error in subsequent gross rock volume estimates. *After Harrell et al. (2004).*

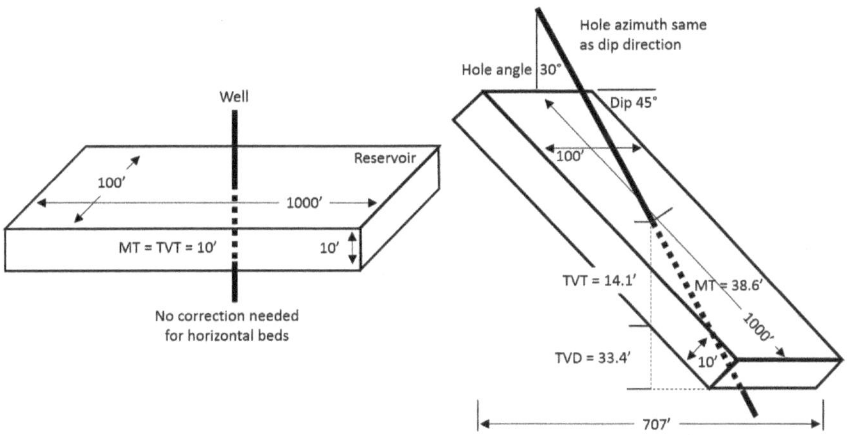


**Figure A1.4** Example of correcting measured thickness (MT) from logs to true vertical thickness (TVT). *After Collarini (2000).*

Mapping of the various reservoir units will also suffer if there is misapplication (or no application) of the NTG factor.

A common mistake is the use of unconstrained and subjective reservoir parameters, which are subsequently input as arbitrary distributions into the probabilistic models employed in volumetric calculations. Errors are also introduced by using averages, weighted averages, or parameter mapping for areal distribution, as these techniques decrease the detail in the reservoir description, making subsequent models of the subsurface too homogeneous. The most appropriate distribution for the GRV parameter is log normal. However, one must beware the impact of long tails when using log normal distributions, as they distort the computed mean value, making it too high. To limit this effect, one should truncate the high-end tail of a log normal distribution to three times the P10 value.

## A1.2.3 Reservoir pressure and temperature

Reservoirs primarily exhibit either normal (hydrostatic) pressure or be overpressured. In the unlikely event that there are no reservoir pressure versus depth data available in the VDR, one can use the value of 0.45 psi/ft with an uncertainty range of ±0.01 psi/ft. To illustrate the variability of this parameter, pressure gradients of up to 0.5 psi/ft have been recorded in the saline formation water reservoirs of the Middle East. The gradients can ramp up significantly to 0.5–0.9 psi/ft when dealing with overpressured regimes such as in high pressure high temperature (HPHT) reservoirs.

Reservoir temperature data are usually provided in the data room but should be cross-checked using the well data from logs and tests. One can estimate the reservoir temperature (Tres) from the relationship:

$$\text{Tres} = (\text{geothermal gradient} \times \text{depth} \div 100) + \text{Ts}$$

where Ts is the seabed or surface temperature if the well is located onshore. It is reported that regional onshore/offshore values of Ts in degrees Fahrenheit are 60/40 which is taken as the standard, 80/50 for the tropics, and 40/30 for the Arctic, respectively. For the record, the geothermal gradient in the North Sea ranges from 15 to 20°F per 1000 ft.

## A1.2.4 Oil formation volume factor

This parameter, denoted as Bo, indicates how much a barrel of oil at reservoir conditions will reduce in volume as it flows to surface and the dissolved solution gas in the oil is released. The units of Bo are reservoir barrel/stock tank barrel or more simply reservoir volume/surface volume. If laboratory PVT data are unavailable, there are numerous black oil correlations in the public domain, but one needs to know regional temperature, pressure, and GOR to use them. GOR has the most significant effect on Bo. A rough and ready estimate of Bo can be obtained from using the relationship $[(\text{GOR}/100 \times 0.05) + 1.05]$.

Alternatively, one can use the following rules of thumb:

- For gassy, volatile oils, with an oil gravity of around 55°API, Bo might be as high as 2 rb/stb.
- For low shrinkage oils, with a likely gravity of 35°API, Bo is likely to decrease to around 1.3 rb/stb.
- Heavy oils, of 15°API or less, which contain little solution gas, may hardly shrink at all, with Bo of around 1.05 rb/stb.

## A1.2.5 Gas expansion factor

Although many of the classic petroleum engineering textbooks use the ratio 1/Bg, where Bg is the gas formation volume factor, many reservoir engineers prefer to use its inverse, termed the gas expansion factor (E). This convention is more representative of what is happening during gas flowing to surface as it expands, rather than shrinks. Also, it avoids the need to use very small decimal numbers in the GIIP calculation.

Being an "expansion" term, the units of E are surface volume divided by the reservoir volume. It is readily determined from pressure, temperature, and gas composition, so it can be mapped after reviewing regional pressure and temperature trends. In the North Sea, the value of E lies typically in the range 240 to 320 scf/rcf, but it may be less than 100 scf/rcf in shallow gas fields and can be almost 400 scf/rcf in deep HPHT fields. For the record, in the United Kingdom, HPHT is formally defined as a reservoir having an undisturbed bottom hole temperature of greater than 300°F (149°C) and a pore pressure gradient of at least 0.8 psi/ft (15.3 lbm/gal) or requiring a BOP with a rating in excess of 10,000 psi (69 MPa).

## A1.2.6 Porosity

The effective porosity is the fraction of the net reservoir that is pore space in which hydrocarbon can be stored. The minimum to maximum ranges of porosity one may experience with sandstone reservoirs ranges from 0.10 to 0.20 for tighter rocks, and 0.20 to 0.35 for more porous rocks. It is accepted practice to characterize porosity with a normal distribution. One should check that net porosity has been clipped on the low side with a minimum value of the chosen cut-off, and on the high side at a theoretical maximum of 0.40, say. The above ranges refer to sandstones and their matrix porosity systems, readers should bear in mind that carbonate reservoirs can have more complex pore geometries which include fractures and vugs.

One simple classification for fractured reservoirs is shown in Fig. A1.5, where reservoir A has high storage capacity in the matrix (around 0.10 to 0.35 porosity), but low storage capacity in the fractures; reservoir B has roughly equal storage in the matrix and fractures; and reservoir C has all storage in fractures.

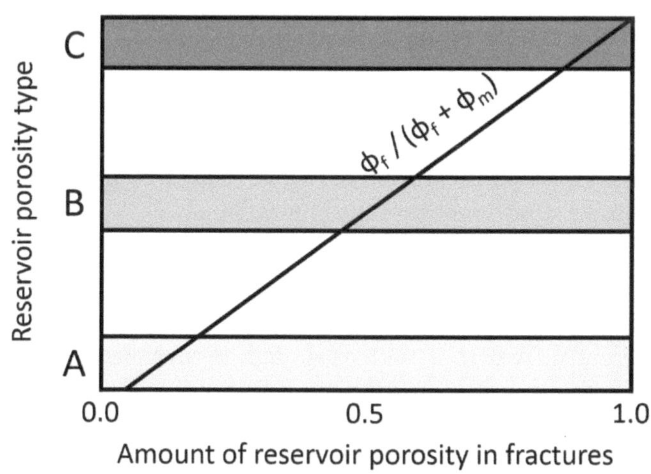


**Figure A1.5** Naturally fractured reservoir classification based on percentage of reservoir porosity contained in fractures. *After Aguilera (1995).*

Matrix porosity ($\phi_m$) is the pore space inherited as primary intergranular or subsequently modified by diagenetic or late cementation or solution effects; it is the porosity generally used in PIIP calculations. Fracture porosity ($\phi_f$) is pore space created by fracturing the rock; it usually contributes a negligible amount to total pore space, but fractures often have high permeability and can contribute substantially to flow. Some ranges of fracture porosity from various geological formations are given below (Weber and Bakker, 1981).

- Monoclines and low-dip anticlines: 0.0001 to 0.001
- Strongly folded anticlines: 0.001 to 0.003
- Fracture porosity enhanced by leaching: 0.002 to 0.01
- Karst aquifers, surface to shallow: 0.002 to 0.03
- Deeply buried brecciated karsts, collapse breccia: 0.005 to 0.02
- Fractured cherts: 0.05 to 0.08
- Fractured tuffs, igneous rocks: 0.02 to 0.08

## A1.2.7 Hydrocarbon saturation

Hydrocarbon saturation is that fraction of the pore space that is occupied by oil or gas in a reservoir. Even if a reservoir is "full" of oil, the hydrocarbon saturation will not be 1.00. This is because in any porous system there is always an irreducible water saturation which has a significant effect on PIIP. The minimum to maximum range of oil saturation is typically 0.60 to 0.80 for tighter rocks, and 0.70 to 0.90 for more porous rocks. These ranges can be reduced by 0.05 and 0.10, for wet gas and dry gas bearing reservoirs, respectively.

The parameter is often characterized with a normal distribution, but some analysts argue that the log normal distribution is more appropriate for describing saturation, as it is more closely related to pore throat size (proportional to permeability), rather than pore size (related to porosity). Yet, as the saturation ranges are usually so narrow, there may be not much difference in the end result. If the range from P90 to P10 is less than the P50 value, then using a normal distribution is acceptable.

In general, saturation height behavior should be included in any PIIP exercise. The only instances where it may not be that important are in "conventional" dry gas fields, where there exists a larger hydrocarbon—water density differential; reservoirs with very high permeability where capillary forces will be lower; and if the petroleum accumulation has high structural relief so that proportionally less of the reservoir is in the transition zone.

### A1.2.8 Net-to-gross

NTG always lies between 0 and 1 and reduces the GRV of a reservoir by the proportion that is non-reservoir (e.g., shale, very tight rock, etc.). Typically, the minimum to maximum ranges of NTG that one may experience can be 0.25 to 0.50 (below average reservoir), 0.40 to 0.70 (average reservoir), 0.55 to 0.85 (above average reservoir), and 0.70 to 1.00 (exceptional reservoirs). As with water saturation, some analysts argue that NTG is better described with the log normal distribution, as the cut-off is based on an arbitrarily chosen permeability value, below which it is believed hydrocarbon will not flow at commercial rates. However, the normal distribution is commonly used to characterize NTG.

## A1.3 Reservoir drive mechanisms and recovery factor in conventional reservoirs

There is a direct linkage between a reservoir's drive mechanism and the expected recovery. The typical drive mechanisms encountered in conventional reservoirs, along with their corresponding minimum to maximum ranges of recovery factors for the different fluid types, are given in Fig. A1.6. After completing their decline curve analyses of production history and generating an estimated ultimate recovery, M&A reservoir engineers should not forget to liaise with their geoscientist colleagues to reality check the corresponding recovery factor against accepted ranges for similar reservoir types, fluids and drive mechanisms.

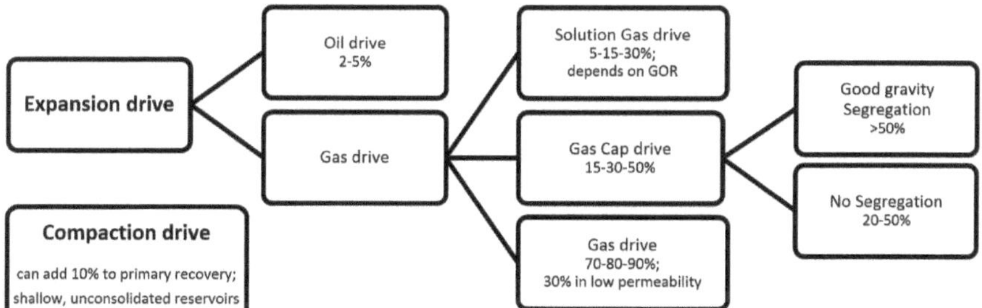


**Figure A1.6** Classification of conventional reservoir-producing mechanisms, with suggested low–mid–high ranges of recovery factor. *After several authors.*

The oil and gas industry defines primary recovery as fluids being produced due to declining reservoir pressure, as opposed to secondary recovery, where the reservoir is repressurized with water and/or gas injection. The types of reservoir energy available for moving hydrocarbons, and the interaction between the fluids and rocks, allows one to classify a reservoir's performance. Tertiary recovery tends to be used when dealing with more viscous oils, by employing thermal techniques and/or gas and chemical injection, to lower the in situ oil viscosity or reduce its interfacial tension, and so increase its mobility and subsequent recovery. The term tertiary recovery is often used interchangeably with enhanced oil recovery (EOR).

## A1.3.1 Water drive

Water drive provides reservoir pressure support and sweeps in situ hydrocarbons to the producing wells and may come from injected water or encroachment from an underlying aquifer. The mechanism can be a mixed blessing in gas reservoirs. Fig. A1.7 shows the typical behavior for a field producing under water drive, with reservoir pressure and GOR staying high due to voidage replacement, and early breakthrough of co-produced water which increases to high volumes.

Water injectors are located downdip of the crestal oil producers, which flow until the water cut becomes excessive, and field economics begin to suffer. There is usually a buildup to peak field oil rate, which is maintained while on plateau due to reservoir voidage replacement, until increasing water cut causes the oil rate to fall. The expected recovery for water drive mechanism can be as poor as 30 percent of STOIIP, but recovery can be over 60 percent in certain serendipitous cases. In the water flooded oil fields of the North Sea, the average recovery is around 45 percent.

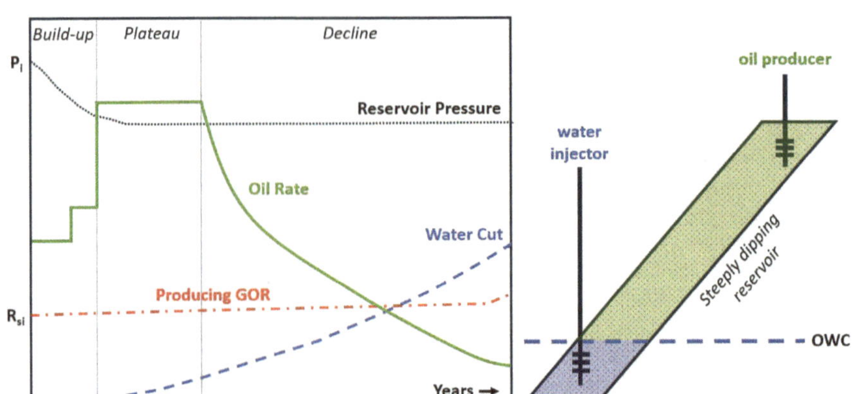


**Figure A1.7** Typical field performance and well locations for water drive. *After several authors.*

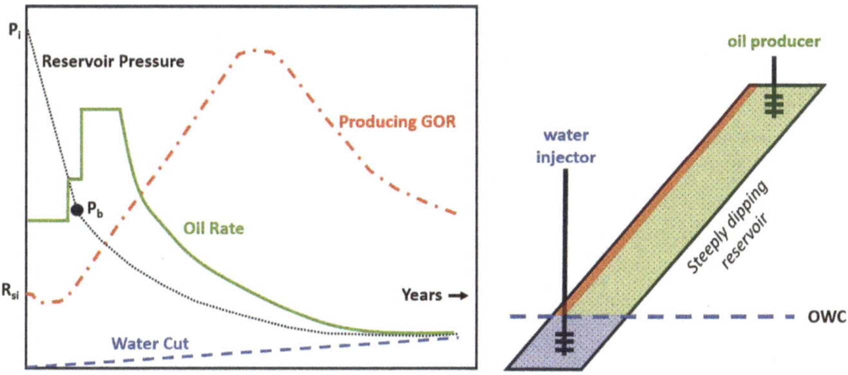


**Figure A1.8** Typical field performance and oil producer locations for solution gas drive. *After several authors.*

## A1.3.2 Solution gas drive

The subdivision of expansion drive depends on whether the oil or gas expansion is the predominant producing mechanism. For example, dry and wet gas reservoirs are classed as having gas expansion drive as there is no free oil at reservoir conditions. Ironically, saturated, black oil and volatile oil reservoirs are described as having a gas drive mechanism. In these reservoirs, the oil contracts rather than expands during pressure depletion due to release of dissolved solution gas. Hence, the vast majority of expansion drive reservoirs are solution gas drive in character, and reservoirs only produce via oil drive while they are undersaturated.

Fig. A1.8 shows the typical behavior for a field producing under solution gas drive, with reservoir pressure in a rapid and continuous decline.

Initially, producing GOR stays low, but rises to a peak after the reservoir pressure goes below bubble point, before reaching a peak and then going into decline. Very little water production is observed in solution gas drive reservoirs. There is no requirement for water injectors. Oil producers are located on the flanks of the structure, away from the crest, and their flow longevity is relative to the size of the created secondary gas cap. The observed field oil production rate will be high initially but will fall rapidly into a continual decline. The expected oil recovery for the solution gas drive mechanism is 5 to 30 percent of STOIIP.

## A1.3.3 Gas cap drive

Besides the solution gas mechanism, gas drive in a reservoir can be due to the expansion of a gas cap, which is a free gas zone that overlies an oil zone that pre-exists or is created during reservoir depletion. Fig. A1.9 shows the typical behavior for a field producing under gas cap drive, with reservoir pressure declining slowly and continuously, whereas GOR rises unceasingly.

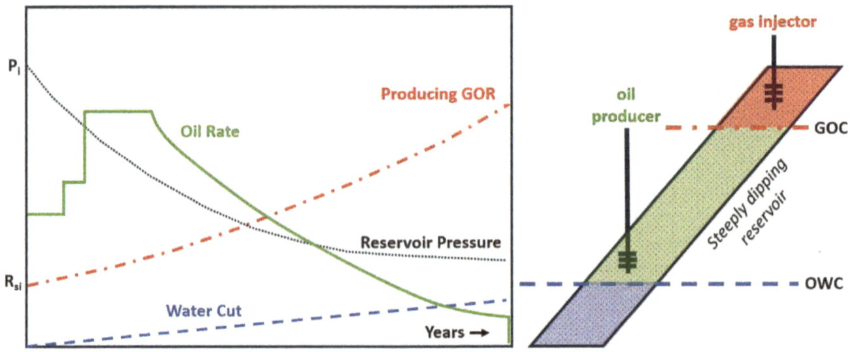


**Figure A1.9** Typical field performance and well locations for gas cap drive. *After several authors.*

Co-produced water is negligible. Gas injectors are located on the crest of the structure, updip of the oil producers on the flanks, whose flow longevity depends on the size of the primary gas cap. Field oil production rate is high at first, but then exhibits a gradual decline. The expected recovery for the gas cap drive mechanism is 15 to 50 percent of STOIIP.

Note that Fig. A1.6 also shows compaction drive, which is far less common than the above recovery mechanisms, occurs when pressure depletion causes the pore volume of the rock to contract and help expel fluids from the reservoir.

## A1.3.4 Identifying reservoir drive mechanisms

If sufficiently detailed field production histories are provided in the data room, potential buyers can plot reservoir pressure and producing GOR against the cumulative oil recovery from a field and use the observed trends to identify which reservoir mechanism is dominant. The theoretical trends for the three main reservoir drive mechanisms are shown in Fig. A1.10.

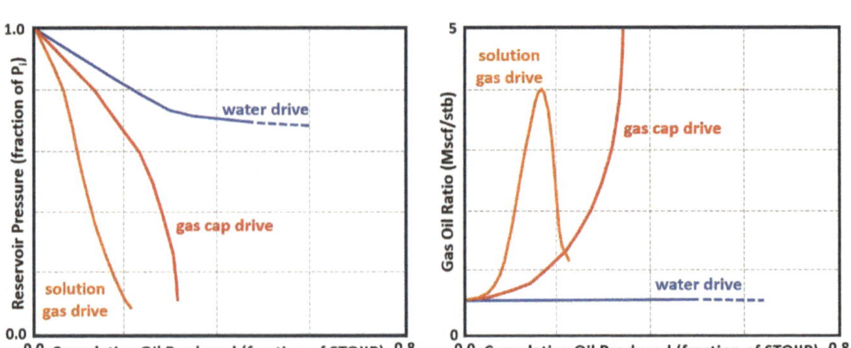


**Figure A1.10** Observed trends in reservoir pressure and gas oil ratio for different drive mechanisms. *After several authors.*

To estimate potential TRVs of petroleum resources, a recovery factor is applied to the estimated PIIP. Hydrocarbon recovery factor is the most important (and difficult) parameter to establish with confidence at any stage of the asset evaluation. It depends on many aspects, both subsurface and at surface, such as the reservoir drive mechanisms, the in situ fluid viscosity and rock properties, the development scheme, operational environment, and the imposed abandonment conditions.

## A1.4 Factors influencing recovery of conventional petroleum resources

If the M&A team is evaluating an exploration asset portfolio, it should only consider potential volumes associated with primary and secondary recovery. Without any wells in the undrilled structure, prospect recovery must be based on local databases to provide analogs, or analytical techniques, or correlations.

The secondary recovery of oil associated with waterflooding can be estimated from three factors: microscopic displacement efficiency, which is the mobile oil saturation; horizontal (areal) sweep efficiency, which is controlled by the shape and extent of the reservoir and by faults; and vertical sweep efficiency, which reflects the stratification in the reservoir. Table A1.1 shows the range of recovery factors that are typically used in the Norwegian sector of the North Sea (NPD, 2017), where the reservoirs are primarily sandstones of medium to good porosity and permeability, containing black oil with reasonable GOR, and have some pressure support from regional aquifers.

However, there are many factors on the effectiveness of a waterflood as a means of secondary recovery, and these are listed in Table A1.2. The

**Table A1.1** Variation in waterflood recovery from microscopic displacement efficiency, vertical displacement efficiency, and horizontal displacement efficiency as observed on the Norwegian Continental Shelf.

| Parameter | Term and equation | Low | Mean | High |
|---|---|---|---|---|
| Initial oil saturation | Soi | 0.80 | 0.80 | 0.80 |
| Residual oil saturation | Sor | 0.30 | 0.20 | 0.10 |
| Microscopic displacement efficiency | Rm=(Soi−Sor) ÷ Soi | 0.625 | 0.75 | 0.875 |
| Vertical displacement efficiency | Rv | 0.7 | 0.8 | 0.8 |
| Horizontal displacement efficiency | Rh | 0.7 | 0.75 | 0.8 |
| Recovery factor | RF = Rm×Rv×Rh | 31% | 45% | 56% |

After NPD (2017).

**Table A1.2** Influences on waterflood effectiveness, causing an increase or a decrease in recovery factor (ΔRF).

| Characteristics of reservoir and fluid | Effect on waterflood recovery | ΔRF |
|---|---|---|
| Poor aquifer permeability | Precludes injection into aquifer | ✗ |
| Poor reservoir permeability | Reduced productivity | ✗ |
| Fault block/faulting | Isolates zone, reduces sweep | ✗ |
| Steep formation dip | Improves sweep by gravity drainage | ✓ |
| Thick reservoir | Increases productivity, reduces coning | ✓ |
| High heterogeneity | Early water breakthrough, poor sweep | ✗ |
| Shallow reservoir (offshore) | Low pressure, reduced areal coverage | ✗ |
| Adverse mobility | Reduces sweep ratio | ✗ |

structure of the reservoir, the flow properties of the rock, both effective and relative, and the viscosity difference between the in place oil and the displacing water all play a role in determining the efficacy of a waterflood. A major takeaway point is that there are far more negative influences than there are positive influences.

If one now considers the factors affecting gas recovery from conventional resource accumulations, it is generally accepted that high permeability reservoirs can recover 80 to 85 percent of the in place dry gas (with compression). At the other end of the spectrum, low permeability reservoirs might recover 40 to 50 percent (with compression). It is vital that the buyer confirms the assumed abandonment pressure for the gas fields under investigation. The pressure at which a gas field will cease commercial production depends on the contract delivery pressure, which is heavily influenced by pipeline pressure drops in the production network. The abandonment pressure can be lower if there is compression available to boost the export pressure of the gas for sale. Based on the normal hydrostatic pressures seem in the gas fields of the UK Southern North Sea and the need to meet the UK onshore grid entry pressure of around 1015 psi (70 bar), a rule of thumb for the region is that abandonment pressure can be estimated from 0.1 psi/ft multiplied by reservoir depth.

Gas field recovery is impaired by water breakthrough at the producing gas wells from an encroaching aquifer, which is exacerbated if the reservoir has highly variable vertical permeability. An insufficient number of development wells leads to poor areal coverage, suboptimal drainage, and ultimately lower gas recovery. Faulting increases the likelihood of compartmentalization and decreases lateral continuity of the reservoir, which means more gas producers will be required to be drilled or recovery expectations will need to be lowered. Secondary cementation, which is caused by diagenesis in the reservoir, is a major problem in some gas fields, especially as it is very difficult to predict which areas may be affected. As its names suggests, secondary cementation reduces reservoir permeability and lowers well deliverability, meaning more producers may need to be drilled to deliver the contractual volumes.

Water drive in a gas field can be a mixed blessing. If the gas accumulation is sizable, with high structural relief and large areal extent, and its reservoir is fairly homogeneous, then the estimated recovery factor may improve by a few percent. However, if the gas accumulation is small and/or stratified, then encroachment due to an underlying aquifer can trap 30 to 50 percent of the GIIP behind the waterfront, which may limit gas recovery to 50 percent at best.

The previous discussions consider "technical recovery" only, but commercial concerns can also impact on how much oil and gas is eventually recovered from a field. For example, the physical location of a petroleum accumulation will determine its ultimate recovery to some degree. If the reservoir to be developed is relatively deep, the associated well costs will be higher and so less producers maybe drilled, with the knock-on effect of lowering recovery. Similarly, the expense of preparing an onshore drilling location may rise if the accumulation is adjacent to urban areas or nature parks, where the costs of land clearance and subsequent reparation are high. Offshore location costs can differ widely depending on the status of the seabed and the water depth. Whether onshore or offshore, the proximity to existing facilities and export routes will dictate the length and cost of infield pipelines for the development. The buyer needs to verify whether extra facilities are required for a particular development. The estimated capital and operating expenditures will inevitably rise if there is a need for accommodation, as it is costly to keep and maintain drilling and production crews in the field. Development costs will also increase if more advanced field processing is needed, such as additional gas treatment to meet stringent export pipeline specifications and potential sulfur removal from oil. Finally, the impact of a country's petroleum legislation and contract terms, which govern export price and a host government's tax take and royalty, can have a significant impact on determining the estimated commercial recovery of petroleum from an accumulation.

# Appendix 2: BOE, a misleading unit

Barrel of oil equivalent (BOE) is used frequently by E&P companies to report the amount of resources they may have and is much beloved by bankers and business development types who use it to compare past M&A deals. It supposedly allows analysts, investors, and management to assess the total amount of energy that companies have access to, without breaking it down into barrels of crude oil or the cubic feet of natural gas.

Converting gas volumes to BOE is traditionally done on the basis of 6 Mscf of natural gas to 1 BOE. The conversion ratio is based on an energy equivalency conversion method, primarily applicable at the burner tip, and does not represent a value equivalency at the wellhead. Therefore, it is obvious that any resources disclosure submitted in terms of BOE is likely to be misleading, particularly if used in isolation, when trying to form a view of a company's value.

## A2.1 Where does the 6:1 rule of thumb come from?

The BOE unit is based on the approximate energy released by burning one barrel of crude oil, which the US tax authorities define as equal to 5.8 MMBtu. The traditional BOE conversion factor is based on the British thermal unit (Btu), which is the heat required to raise the temperature of 1 pound of water by 1°F, as even though gas production is measured in volumes, gas price is still quoted in \$/MMBtu. However, the quoted calorific values (CV) of petroleum are approximate, as various grades of oil have slightly different heating values and sales gas can vary in quality at the point of sale, which equates to the Reference Point where reserves are estimated using the PRMS.

Table A2.1 shows the reported conversion factors used by some of the major oil companies, which reflects a spread of 10% and demonstrates that the difference in total reserves by using different conversion factors can be large.

According to the US Energy Information Administration (EIA), in 2016–18, the average heat content of natural gas sold in the United States was about 1037 Btu/scf, which gives 1 Mscf equal to 1.037 MMBtu. This compares to the UK Government's stated CV of metered sales gas of 1.02264 MMBtu/Mscf. Both of which are lower than the EIA's stated CV

**Table A2.1** Variation in factors to convert from MMscf to BOE used by major oil companies.

| Major Oil Company | MMscf to BOE conversion factor |
|---|---|
| ExxonMobil | 6.000 |
| ConocoPhillips | 6.000 |
| Chevron | 6.000 |
| Shell | 5.800 |
| BP | 5.800 |
| Repsol | 5.615 |
| Statoil | 5.615 |
| ENI | 5.492 |
| TOTAL | 5.390 |

Spread equates to 10% difference in conversion factor

After Morales (2016).

conversion for liquefied natural gas (LNG), which has a much richer composition, of 1.100 MMbtu/Mscf.

Should investors be concerned? Yes, they should. Many analysts believe the conversion formula should be based on price received rather than an outdated volumetric measurement of heat produced.

## A2.2 Local rule, not global standard

There is also the issue of where an oil and gas company's assets are located and what markets they serve which needs to be taken into consideration. Table A2.2 shows the variation and fluctuations in BOE conversion based on oil and gas prices for the past 5 years for several of the major global hubs.

Consider that the 2018 average hub prices for the US benchmarks of West Texas Intermediate crude at Cushing, Oklahoma and natural gas at Henry, Louisiana, which were \$64.8/stb and \$3.1/Mscf, respectively. This gives a BOE conversion ratio (based on product price) of $64.8 \div 3.1 = 20.9$ Mscf of gas to 1 stock tank barrel (stb) of crude. Hence, the traditional BOE conversion by thermal energy would be in error by $20.9 \div 6 = 3.5$ and would significantly overvalue the reserves stated in any disclosure by 350%!

Now consider Europe, where the 2018 average spot prices for North Sea Brent crude and European Union Border imported gas were \$71.1/stb and \$7.5/Mscf, respectively. The traditional BOE conversion would be in error by a factor of 1.6 and overvalue the reserves claimed in the disclosure.

**Table A2.2** Difference in BOE conversion factors, based on oil and gas prices in different global regions versus traditional BOE conversion factor of 6 Mscf/barrel.

| | Location | 2013 | 2014 | 2015 | 2016 | 2017 | 2018 |
|---|---|---|---|---|---|---|---|
| \$/Mscf | Europe | 11.5 | 9.8 | 7.1 | 4.5 | 5.6 | 7.5 |
| | USA | 3.6 | 4.2 | 2.5 | 2.4 | 2.9 | 3.1 |
| | Japan | 14.5 | 14.6 | 9.5 | 6.7 | 7.8 | 9.7 |
| \$/barrel | Brent | 108.6 | 99.0 | 52.3 | 44.0 | 54.4 | 71.1 |
| | WTI | 98.0 | 93.2 | 48.7 | 43.3 | 50.9 | 64.8 |
| | Dubai | 105.4 | 96.7 | 51.2 | 41.2 | 53.1 | 69.2 |
| BOE based on price (\$/barrel divided by \$/Mscf) | Europe | 9.4 | 10.1 | 7.4 | 9.9 | 9.7 | 9.5 |
| | USA | 27.0 | 22.0 | 19.3 | 18.0 | 17.8 | 21.2 |
| | Japan | 7.3 | 6.6 | 5.4 | 6.1 | 6.8 | 7.1 |
| Difference in BOE conversion; price basis versus heat equivalence | Europe | 157% | 168% | 123% | 164% | 162% | 158% |
| | USA | 450% | 367% | 321% | 300% | 297% | 354% |
| | Japan | 121% | 110% | 90% | 102% | 113% | 119% |

Data from EIA and World Bank.
WTI = West Texas Intermediate.

The situation is not as bad as in the United States, but using BOE conversion still overvalues European gas assets by nearly two thirds.

Yet if one now considers the Asia Pacific region, the 2018 average spot prices for Dubai crude and Asia LNG were \$69.2/stb and \$9.7/Mscf, respectively. This provides a BOE conversion ratio (on a price basis) which is closer to the traditional thermal energy equivalent figure of 6. This is not surprising, as the LNG contracts in Japan, Korea, Taiwan, and China are indexed traditionally to the prevailing oil price. However, this relationship may be weakening as there is a reported trend of breaking the link between LNG prices and crude oil prices in the region.

The important take away point is that the use of BOE will overvalue gas assets wherever they are located in the world.

## A2.3 Gas composition is important

The traditional rule of thumb for the BOE conversion assumes that gas compositions are known, which is usually not the case when dealing with prospective resources. The conversion factors that tend to be quoted are for fairly lean natural gas with low inert content. High levels of $CO_2$ will drastically lower the CV of the gas and so alter the conversion factor, as it is energy based.

## A2.4 The 50% rule to decide whether to report cfge or BOE

To allow an oil and gas company's corporate health to be gauged more meaningfully, there is a simple fudge that investors can use to bring clarity; it is known as the 50% rule.

If a company has an asset portfolio with over 50% gas resources, it is probably more appropriate to convert oil volumes into a gas equivalent in terms of cubic feet of gas equivalent (cfge). By the same token, if a company has an asset portfolio with over 50% oil resources, it is probably more appropriate to convert gas volumes into an oil equivalent in terms of BOE.

For example, consider an oil and gas company that has 480 Bscf of natural gas reserves and 20 MMstb of oil reserves. If the company reports its reserves on a BOE basis, using the traditional conversion ratio of 6:1 to convert its gas reserves, then it may submit a figure of

480 ÷ 6 converted gas reserves + 20 MMstb oil reserves = 100 MMBOE

If then one applies the 50% rule, 20 MMstb ÷ 100 MMBOE = 20%, which is far less than 50%.

Alternatively, if the company quoted its resources in cfge by reporting a figure of

480 Bscf gas reserves + 20 × 6 converted oil reserves = 600 Bcfge

If then one applies the 50% rule, 480 Bscf ÷ 600 Bcfge = 80%, which is a much better result.

The computed equivalent oil volume in the above example is five times greater than its actual oil reserves. However, if the 50% rule is applied and the oil and gas company's reserves are reported in cfge, it would help investors to see clearly and quickly that the company is primarily a gas-rich operator with resources of 600 Bcfge, which is much closer to the actual resource makeup of the company.

If companies with primarily gassy portfolios report their resources using BOE, then this should raise a red flag to potential investors.

# Appendix 3: Estimating exploration and development risks

One of the features of the oil and gas sector is the high level of risk associated with the investment process at nearly all stages of the business cycle. Arguably, the two principal hurdles that an asset faces along its journey, from prospect to discovery to producing field, are those of deciding to drill the exploration well and, in the event of finding producible petroleum, electing to proceed with its development. The risks associated with these hurdles are termed the chance of discovery, which is also known as the geological chance of success, and the chance of development, which is the likelihood that the asset will attain commercial producing status.

The M&A team must be reassured that the estimated risks associated with exploration and development (which the seller should provide) are appropriate for the assets under review. There are many components of these two principal risk categories, which not only deal with subsurface issues but also with technical, commercial, environmental, and political factors. These elements are covered in this appendix.

Overcoming the exploration and development risks is fundamental to the resource categorization within the SPE's Petroleum Resources Management System (PRMS), as shown in Fig. A3.1. Prospective resources have a chance of discovery ($P_g$) and a chance of development ($P_d$), whose product is the chance of commerciality ($P_c$). If the exploration well is a success, the chance of discovery becomes 100 percent, so contingent resources have only to consider a chance of development.

Let us take a closer look at the chance of discovery.

## A3.1 Chance of discovery

The PRMS has three subclassifications of prospective resources; play, lead, and prospect, which reflect the increasing knowledge and understanding of what hydrocarbons, both volume and type, may be contained therein.

Stoneley (1995) defines a play as a group of hydrocarbon fields, discoveries, or prospects that are located in the same region, and being controlled by the same set of geological circumstances. Sellers will claim to have quantified the total potential of a play, and an example of the results of a

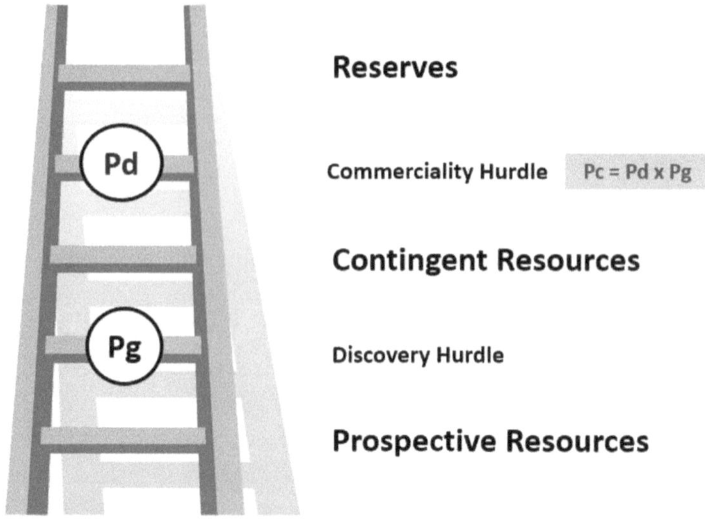


**Figure A3.1** Moving up the PRMS classification "ladder" by overcoming discovery and development or commerciality hurdles. *After SPE (2018).*

seller's play resource assessment is shown in Table A3.1. Such an assessment requires benchmarking to actual and analogous field data to indicate likely average trap size, plus some idea of trap density. These play assessments provide a broad indication of potential only, and the seller's numbers should be treated with caution by the M&A team (Scallon, 2015).

As a play is worked up, so leads and prospects are identified; a prospect is ready to be drilled, whereas a lead is not. When assessing a lead, its maturity will depend on how close it is to being an upgraded to a prospect. If the lead is reasonably well defined, the seller should give a range of low/best/high resource estimates, as is done for a prospect, but if it is poorly constrained, then the seller should consider providing a single indicative best estimate. Prospect assessment is the same as for a lead, but the structural risk component should be less, as more investigative work will have been performed.

**Table A3.1** Example of a simple deterministic exploration play assessment.

| Play | Low | Best | High | Risk |
|---|---|---|---|---|
| Assumed number of traps | 3 | 8 | 15 | High |
| Mean trap size (MMstb) | 60 | 60 | 60 | High |
| Total potential resource (MMstb) | 180 | 480 | 900 | High |

After Scallon (2015).

When it comes to checking the seller's exploration asset portfolio, the buyer's M&A team need to focus on the seller's low case assumptions and input parameters. These usually receive the least attention, but they can have significant effect on the ranking of the opportunity, and impact on the associated best and high cases if working with an assigned log normal distribution.

### A3.1.1 Elements of the likelihood of geological success

There are several models used in the industry to estimate the chance of a discovery or the geological chance of success (Pg) for an exploration well. One of the classic models is that of Otis and Schneidermann (1997), which estimates the chance of a discovery from the product of the individual probabilities of occurrence for four elements:

$$P_g = P_{Source} \times P_{Dynamics} \times P_{Trap} \times P_{Reservoir}$$

All of the probability elements must be present for any chance of the prospect containing hydrocarbons. Attempts to use any relative weighting of the probability elements are irrelevant, as the absence of any one of them will invalidate the prospect.

Let us take a look at each of the probability elements in turn, with a view of what the M&A team should be looking for.

$P_{Source}$ is the probability that a mature source rock exists and has generated enough hydrocarbon to charge the reservoir. It is dependent on source presence and effectiveness. The M&A team must confirm that the seller has demonstrated that the source rock is present in the mapped prospect "kitchen," via records of hydrocarbon shows and seeps, and has identified the source rock facies in offset wells and outcrops in the area. Also, the seller should provide an estimate of source rock volume and, based on total organic content and kerogen type, the seller should propose whether the source rock is oil or gas prone.

$P_{Dynamics}$ is the probability that the timing of generation of petroleum was compatible with the formation of the trap, and that a viable migration path, from source to reservoir, existed during that time window. To assign source and dynamic risks, the seller must also have investigated the relative timing of expulsion and trap formation and estimated the generation potential of the source kitchen with respect to the suggested prospect size. The M&A team should get assurance that the seller has investigated migration efficiency and has documented possible migratory pathways required to charge the traps and prospects.

$P_{Trap}$ is the probability that a closed and sealed structure, or stratigraphic trapping feature, exists within which hydrocarbons may accumulate. In estimating its presence and effectiveness, the seller should have considered likely trapping geometries and mechanisms, and reviewed the burial history of the area, its structural trends, and evidence of fault sealing. The M&A team should confirm that the seismic data acquired over the seller's exploration acreage are sufficient in coverage, quality, and resolution to identify the trap. The seller's major seismic horizon picks and the quality of the well ties should be checked. If the mapping is not performed at the top reservoir pick, one must ascertain what surface was used and how projections were made to the reservoir level to create the structure map. Mapped faults and contouring should be checked for realism. Regarding the trap's effectiveness, the M&A team should confirm the lithology of trap seals and determine whether this formation acts as a seal elsewhere. One should review the seller's modeled fault throws to gauge whether they would help trapping. Also, one should cross-check that the claimed analog traps work elsewhere in the basin under scrutiny.

$P_{Reservoir}$ is the probability that reservoir quality rock (i.e., rock that is porous and permeable) exists within the trap. Porosity is needed to provide storage for hydrocarbons migrating into the trap; permeability is required for productivity to allow the hydrocarbons to flow into the wellbore. Estimates of potential values of these parameters are based on diagenetic history, sedimentological studies, porosity types observed in such reservoirs, and likely reservoir parameter variability. To check the seller's view of reservoir presence, one should confirm that the reservoir has been identified and mapped from seismic, offset wells, and outcrops. The seller's proposed reservoir parameters should be reviewed and checked for compatibility with those expected in the seller's assumed depositional environment and facies model. The M&A team should consider how the seller's selected facies model may impact on the geometry and estimated rock volume of the reservoir in the trap. Regarding the effectiveness of the reservoir, the seller should be able to demonstrate that he has exhibited stable flow elsewhere in the play. The M&A team should confirm how the distribution and variation of porosity and permeability have been modeled by the seller, and whether there is any diagenesis overprint.

Some geoscientists argue that it is sufficient to lower the probabilities associated with the hydrocarbon type that might be found in the prospect, i.e., those that deal with source and timing, to account for the uncertainty of whether oil or gas will be present. The author disputes this practice of

**Table A3.2** Generic range of assigned geological chance of success, depending on geoscientific knowledge and data quality and quantity.

| Geological chance of success | Assigned Pg | Geoscience control |
|---|---|---|
| Very likely | 0.9 (1 in 1.1) | Very good, prospect data favorable |
| Likely | 0.7 (1 in 1.4) | Ambiguous, but analogs unambiguous |
| As often as not | 0.5 (1 in 2) | Limited or no data, case not proven |
| Sometimes | 0.3 (1 in 3.3) | Ambiguous, analogs unfavorable |
| Very unlikely | 0.1 (1 in 10) | Very poor, inappropriate analogs |

fudging the "hydrocarbon risk" as it is not transparent to decision makers as to what outcomes may occur if the wildcat be drilled. It also makes it difficult to audit what was done and why certain risk values were assigned if a dry hole postmortem is subsequently carried out. This is the background to the case study described in Section 9.2.

If the potential buyer is comfortable that the seller has a sound geoscientific understanding of the area to be explored, there is a reasonable chance that some of the prospects on offer will find hydrocarbons. However, besides extensive coverage of good quality 3D seismic data, and nearby well data, it is fair to say that many major hydrocarbon discoveries needed a slice of luck. The limits of geological risking are shown in Table A3.2.

Although the figures provided in Table A3.2 look promising, explorers tend to operate in the lowermost regions of the "sometimes" and "very unlikely" categories for the chance of discovery. Even in mature basins in the UK North Sea, the historical value of geological success for a Northern North Sea oil prospect is still only 0.12 to 0.14 (1 in 7–8). However, a Southern North Sea gas prospect is more likely to be discovered with a historical value of geological success of 0.33–0.5 (1 in 2–3), whereas prospects in frontier areas like the West of Shetland are assigned a much lower value of around 0.066 to 0.1 (1 in 10–15).

Let us turn our attention to the chance of development.

## A3.2 Chance of development

The limited guidance given in the PRMS on the estimation of chance of development is not prescriptive, and merely indicates that all contingencies that prevent commercial development should be fully described, and that these may be quantified, without instructing users how it can be done.

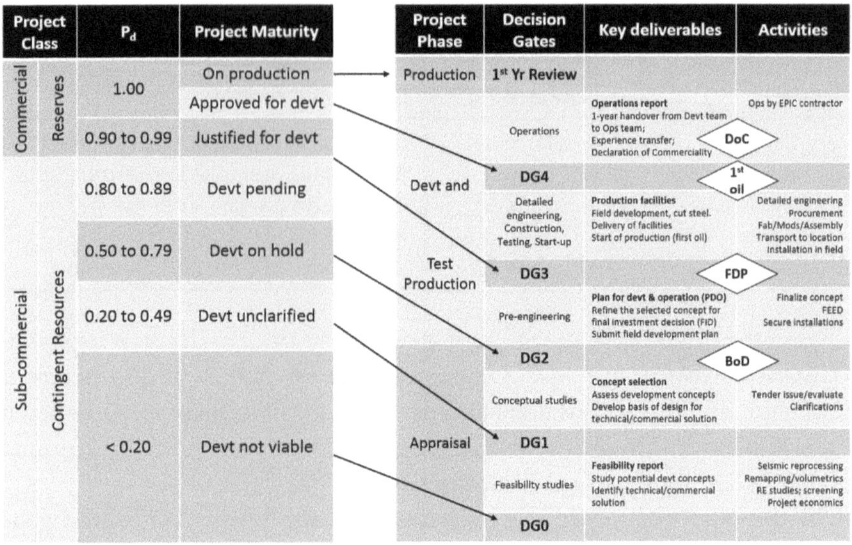


**Figure A3.2** PRMS grid showing resource classifications and subclasses, with suggested chance of development, and suggested links to a decision gate process.

However, a quantitative estimate of the chance of development, usually expressed as a percentage, is specified as a requirement when reporting contingent resources under some stock exchange reporting regulations.

As many resource reporting systems are project-based, it makes sense to link the chance of development estimates with defined development plans, timeframes, and related cost and production profiles for the contingent resources. This approach should reflect the corporate decision gates that a project must pass as it transitions to commercial production, and define its appropriate subclass of contingent resources, as shown in Fig. A3.2.

Currently, there is no widely accepted methodology to guess the chance of development, but a consistent framework for its estimation would be an improvement on the subjective nature of many such estimates employed today, and their use in commercial transactions and regulatory disclosures. The proposed estimating model needs both project-specific and extraneous contingencies to be considered. The various elements of the chance of development model are discussed in the next section.

## A3.2.1 Elements of the likelihood of commercial production

The suggested model for estimating the chance of development for a project captures the various contingencies within four elements: economic viability, technical understanding and reliability, field development plan (FDP) status, and any other contingencies that an Operator may be unable to influence (after Braaten, 2015). The chance of development ($P_d$) is calculated as the product of the probability of success of each of these four elements:

$$P_d = P_{Economics} \times P_{Technology} \times P_{FDP} \times P_{Other}$$

Let us review each of the model's probability elements in turn and highlight what the M&A team should be looking for.

$P_{Economics}$ will be equal to 1.0 if the project is economic under the same fiscal conditions as required for reserves. Typically, a net present value at a hurdle discount rate of 10 percent at a given reporting date (NPV10) is regarded as the economic pass mark, but companies may decide to use their weighted average cost of capital (WACC) instead. This WACC rate reflects the company's gearing or split of debt and equity in its corporate capital structure, which dictates its cost of borrowing. Any project under consideration needs to deliver a return in excess of WACC so that the developer can repay his lenders and issue dividends to his shareholders. Economic contingencies include uncertainty of future hydrocarbon price forecasts, cost trends in the region as oil and gas projects usually involve long lead and construction times. There may also be concerns about the status of ongoing or future contract negotiations, possible changes to taxation, royalty or production sharing contract terms, and adverse exchange rate movements. Entitlement to resources is a constant worry, such as when an operator's production license is approaching its expiry date and there is uncertainty over its renewal and whether the terms will be as generous as before.

$P_{Technology}$ may be assigned a value of 1.0 if the reservoir performance is well understood and the designed project uses established reliable technology. Yet there may be many technical risks to overcome, such as poorer than expected well productivity, premature water breakthrough due to unexpected permeability contrasts, and facilities' hardware failure, e.g., gas compressor malfunction, which may lead to the field being shut in if flaring is disallowed. There are also subsurface risks one may encounter during development drilling when it may transpire that the structural interpretation of the discovery is incorrect, leading to dry development wells. Also, it may become evident that there is more faulting which can compartmentalize the reservoir so that more wells are required to drain it.

$P_{FDP}$ may be assigned a value of 1.0 if a project has a fully worked up and costed development plan that has the required approvals to go ahead. If the FDP provided by the seller is more conceptual, with less detail and more uncertain cost estimates, and the necessary approvals have yet to be gained, then the value of this element will be less than 1.0. The timing of the development has a huge influence on this risk element, as unforeseen challenges may arise in projects with longer lead times. The M&A team must be satisfied that the seller has a track record that shows he is able to execute a project, or multiple projects, in a given timeframe. Some due diligence concerns are the seller's commitment to proceed, whether there are regulatory and partner approvals to be obtained, the infrastructure capacity and reliability for export, and the availability and access to markets. The impact of these factors varies from project to project but is especially important for gas projects as they move toward commercial development.

$P_{Other}$ is the element that considers the risks associated with non-technical issues that are not previously addressed. These contingencies, which are usually political, social, and environmental in nature, are known to the seller, but are often outside his control. The due diligence should consider possible changes in a host government's energy strategy, such as nationalization of assets, or changes to the country's conservation and environmental policies, or the threat of social unrest and the potential for war breaking out, which would impact field operations and possibly incur international sanctions.

## A3.2.2 Example of estimating the chance of development

Consider an E&P start-up which has made an offshore gas discovery in the Asia—Pacific region. The estimated contingent resource volumes are significant, but the gas does have a high $CO_2$ content, which impacts on the facilities required, the export routes available, and the subsequent sales price obtained. The proposed development concept is considered viable, but further appraisal drilling is needed, and additional studies are required to reduce cost uncertainty and highlight the key risks.

The fledgling company believes a standalone floating production storage offshore (FPSO) option for the development, designed specifically for the discovery, would mitigate many of the potential technical risks, as solutions can be incorporated into the design of the vessel.

There is some anecdotal evidence that an export route via existing infrastructure in the region is planned to be installed, but there are no firm indications of timing, which means there is a high risk that the export route may not be available to meet planned first gas production in a few years' time.

A further concern is that the development concept is dependent on a third party gas sales agreement (GSA) being in place, as well as installing and funding the downstream infrastructure; all of which are out of the direct control of the E&P start-up. However, the host government, which is keen to see more external investment in the country's gas resources, has encouraged the start-up to consider progressing the development.

From the identified contingencies outlined above, the estimated chance of commercial development occurring within a reasonable timeframe for the gas discovery is calculated below.

$$P_d = P_{Economics} \times P_{Technology} \times P_{FDP} \times P_{Other}$$

$$P_d = 0.90 \times 0.80 \times 0.65 \times 1.00 = 0.47$$

With reference to Fig. A3.2, the perceived risk-associated project suggests it should be described as contingent resource subclass, "Development Unclarified,"

## A3.3 Avoid using chance of development to aggregate resources

When buyers strive to arrive at a bid that will satisfy the seller, there is usually a need to aggregate the different resources classes in the asset portfolio on offer, which means having to assign commercial values to the reserves, contingent resources, and prospective resources.

The problem arises that resource reporting systems oppose the aggregation over resource classes, as the various classes of resource volumes may not be ultimately produced, as they depend on the success of the project. The due diligence assessment, which is inherently conservative, audited, and which has to be compliant with a stated system such as PRMS, are usually lower than that of the seller, whose in-house valuation is invariably more optimistic, usually unaudited, and tends to be non-compliant with PRMS. The seller will want some value attributed to those resources that are not classified as reserves, but how much value should be assigned by the potential buyer is a highly subjective matter.

The PRMS application guide (2011) states that "*it is important not to aggregate Reserves, Contingent Resources, and Prospective Resources without due consideration of the significant differences in the criteria associated with the classification that comprise the risk of accumulations not achieving commercial production.*" Therefore, PRMS strongly recommends that the resource class components are identified separately and not reported as a single aggregate number.

However, buyers do need a single number to formulate a bid.

The PRMS recognizes that an asset portfolio is evaluated by some in the oil and gas industry using "risked volumes," which are calculated by multiplying mean success volumes by the probability of success, which include the chance of discovery and the chance of commercial development. Indeed, the author has observed buyers multiplying the 2C and Prospective Mean estimates by their respective estimated chances of commerciality, and then adding the resultant risked values to that of the 2P reserves. This practice is frowned upon as it mixes volumes of different risk and of different confidence intervals.

To stay within the PRMS guidelines, auditors need to identify, classify, evaluate, and report each of the various resource components. It is permitted to "stack" these volumes to provide "an idea of the potential" of the future projects and the asset portfolio. Yet, it is imperative that buyers do not lose sight of the fact that these stacked volumes are a mix of projects with different chances of fruition, which means the likelihood of the stacked volume curve occurring is highly uncertain. Auditors should point this out clearly to every potential buyer.

Both aggregation processes described above will produce completely subjective values, as adding up such volumes results in a "statistical expectation of the recovery," as one is mixing deterministic and probabilistic methods, which rarely reconcile with each other. However, using them as a potential upside in conjunction with the 2P reserves estimate, which serves as a baseline, will provide buyers with a ball park estimate of where a successful bid might lie.

# Appendix 4: Resource reporting systems that meet international standards

This appendix provides and overview and comparison of the systems for reporting hydrocarbon resources that are regarded both as international standards and fit for use by the global financial markets.

When hydrocarbon resources can be technically and commercially exploited, they are called reserves. The amount of reserves of an entity shows the potential for commercial growth, allowing it to raise funds in a publicly traded scenario. Yet, there is neither a clear objective and agreed way to calculate them nor is there any unique legislation for certifying reserves (Câmara et al., 2015).

Externally reported resource data are used by governments and their regulatory agencies for national reserves reporting and to direct energy policy, and by banks to support project finance arrangements. All stakeholders, whether they are oil and gas companies, regulators, governments, banks, or investors, require complete, consistent, and reliable information on future production and associated cash flow estimates through full life recovery (Seager, 2013).

In principle, the classification of resources requires a common framework for estimating quantities of oil and gas, both discovered and undiscovered, associated with reservoirs, properties, and projects. Embodied in a code that should cover volumes of petroleum originally in place, technically and commercially recoverable, on production or produced. To be regarded as an international standard, the code should give clear terms and definitions to establish a universal language for reliable and comparable resource estimates (SPE, 2018).

## A4.1 Stock exchange requirements for resource reporting

All oil and gas companies that are registered with stock exchanges must submit annual statements that are compliant with the guidelines laid out in the bourses' selected resource reporting system. Although the bodies responsible for these systems and the stock exchanges who use them share agreed definitions of resources, the systems provide a framework and guidance on their application, while the stock exchanges stipulate reporting requirements, such as frequency and reconciliation. In most cases, the resource reports that are submitted to a stock market must be verified and signed off by a Competent Person, external to the filer, who must work for a recognized resource auditing company, and who must provide an independent and unbiased view of the assets (Scott, 2012). In contrast to stock exchanges, when reporting at a national level, government agencies attempt to capture the full resource base to forecast future production potential for the country. The information regarding production and reserves is used to implement and modify legislation and policy on resource development to manage energy supply. There are several resource reporting systems in use around the world, as evinced by Table A4.1.

The guidelines that are used by most of the bourses that trade in oil and gas stocks include the PRMS; the Canadian Oil & Gas Evaluation Handbook (COGEH), written by the Society of Petroleum Evaluation Engineers (SPEE) Calgary Chapter in collaboration with the Canadian Securities Administrators (CSA); those of the Norwegian Petroleum Directorate (NPD); and the US system regulated by the Securities and Exchange Commission (SEC). Table A4.1 gives an estimate of the number of each resource reporting system that is used by the listed oil and gas companies around the world, and the PRMS is clearly the most popular system in use today. Apart from the SEC, the other three systems share a common heritage, stemming from the resource classification work done in the period 1987 to 2000, by the SPE, the World Petroleum Congress (WPC) and the American Association of Petroleum Geologists (AAPG).

Let us look at the resource reporting systems more closely.

**Table A4.1** Stock exchanges and their preferred resource reporting systems (HKeX, 2009; PwC, 2013; SGX, 2013; SCM, 2017; Reeder, 2017).

| Stock Exchange | Country | No. of listed companies | Listed oil & gas, energy services companies | Percentage Representation | COGEH 2009 | NPD 2016 | PRMS 2007 | SEC 2010 |
|---|---|---|---|---|---|---|---|---|
| TSX/TSXV | Canada | 3,164 | 315 | 10.0% | ✓ | | | |
| NYSE/NYSE | USA | 2,316 | 207 | 8.9% | ✓ | ✓ | | ✓ |
| NASDAQ | USA | 2,872 | 53 | 1.8% | ✓ | ✓ | | ✓ |
| LSE/AIM | UK | 2,614 | 170 | 6.5% | ✓ | ✓ | ✓ | ✓ |
| ASX | Australia | 2,066 | 136 | 6.6% | | | ✓ | |
| HKeX | China | 1,930 | 29 | 1.5% | ✓ | | ✓ | |
| Oslo Børs | Norway | 215 | 53 | 24.7% | ✓ | ✓ | ✓ | ✓ |
| SCM | Malaysia | 986 | 70 | estimate 5-10% | ✓ | | ✓ | |
| SGX | Singapore | 757 | 30 | 4.0% | | | ✓ | |
| Total Companies | | | 1063 | | 335 | 53 | 407 | 269 |
| % of Companies | | | 100% | | 32% | 5% | 38% | 25% |

## A4.2 PRMS

The SPE recognized the need for common standards for petroleum resource definitions to provide consistency, transparency, reliability, and set about creating and maintaining an international standard petroleum reserves and resources classification system based on industry best practices. After many years of development, the PRMS was first published in 2007 and was originally intended to improve the consistency of internal resource assessment and so help oil and gas companies to better manage their E&P businesses. To this end, the PRMS′ classification framework and definitions are the current technical standard for most oil and gas companies, who have adapted PRMS into their own internal systems to report recoverable volumes and value. However, this "adaptation" means that the PRMS understanding used in one company may differ from that used in another company.

The PRMS is project-based, using a base case comprised of the resource evaluator's forecast of future conditions, and it can be applied to both conventional and unconventional resources. The guidelines are principles-based, which gives users the flexibility to tailor the system to their needs but can often cause problems for resource auditors as sellers often fail to apply the PRMS guidance with sufficient rigor. It is regulator-neutral as it does not modify the interpretation or application of any existing regulatory reporting requirements, which makes it easier for oil and gas companies to use. Even though the PRMS was not designed to be a system for public disclosure, that is precisely what it has become, via its adoption by the stock exchanges in London and the Asia—Pacific region.

The PRMS is a global framework, which is used for national resource management and reporting in several countries, including Argentina, Bolivia, Brazil, Colombia, France, Italy, Mexico, Netherlands, South Africa, Uganda, United Kingdom, United States (offshore), and is being considered by Kazakhstan. The system uses a cradle to grave framework that categorizes resources according to the degree of uncertainty associated with their recoverable volumes (Fig. A4.1, horizontal axis) and classifies resources according to their potential to attain commercial producing status (Fig. A4.1, vertical axis).

The goal of the PRMS is to move the projects up the ladder from the prospective resources, through contingent resources and into reserves. Although the PRMS only mentions it briefly, reserves can be downgraded and reclassified as contingent resources if the operational and commercial climate changes for the worse (Harrison and Falcone, 2015). For example, downgrading can occur more readily in the Netherlands, which was one of

| | | | | |
|---|---|---|---|---|
| TOTAL PIIP | DISCOVERED PIIP | | PRODUCTION | Project Maturity sub-classes |
| | | COMMERCIAL | RESERVES<br>1P - 2P - 3P | On production |
| | | | | Approved for devt |
| | | | | Justified for devt |
| | | SUB-COMMERCIAL | CONTINGENT RESOURCES<br>1C - 2C - 3C | Devt pending |
| | | | | Devt on hold |
| | | | | Devt unclarified |
| | | | | Devt not viable |
| | | UNRECOVERABLE | | |
| | UNDISCOVERED PIIP | | PROSPECTIVE RESOURCES<br>1U - 2U - 3U | Prospect |
| | | | | Lead |
| | | | | Play |
| | | | UNRECOVERABLE | |

**Figure A4.1** The common resource classification framework used by PRMS. The differences between initial and updated versions is stated to be minimal, but some important changes have been made. *After SPE (2018) and Carroll (1871).*

the first countries to adopt PRMS and uses the system pro-actively to manage its national petroleum resources. Quite sensibly, the Dutch regulator reduced the length of time that reserves can languish in the "Justified for Development" subclass of reserves (see Fig. A4.1) to 2 years after which the resources are downgraded to contingent resources.

A set of application guidelines to accompany the 2007 version of PRMS were published in 2011. An updated version of PRMS was released in 2018 with changes that were stated to be "evolutionary rather than revolutionary", being a Tweedle Dum and Tweedle Dee situation with the new system being very similar to the original. Certainly, the subjective language and lack of prescription remain, as does its dubious handling of decommissioning (Harrison and Falcone, 2017), but there are some changes that bear closer scrutiny. For example, the move to allow fuel gas to be included in reserves should make buyers and auditors beware. Also, by now allowing different projects to be defined for the low, best, and high cases for contingent and prospective resources, the PRMS appears to be mixing risk with uncertainty in the X-axis of the framework and would seem to advocate split conditions, which the system categorically states are not allowed.

## Curiouser and curiouser

There was an opportunity for the SPE and SPEE to work together and merge the PRMS and COGEH codes, but the societies failed to put aside their differences and the codes were updated independently from each other in 2018. This is a pity, as a unified reporting system, taking the best from either code, would have better served the industry, institutions, investors and governments alike. Let us now consider the COGEH reporting system.

## A4.3 COGEH

The Canada Petroleum Resources Act forces a company responsible for discovering hydrocarbons to ask the National Energy Board (NEB) to declare either a "significant discovery" to suggest the existence of a potential hydrocarbon accumulation to be produced or a declaration of "commercial discovery," which shows there is enough oil to justify the investment to produce it. Using the "Joint Guidelines Regarding Applications for Significant or Commercial Discovery Declarations and Amendments," a document of the Canada–Newfoundland Offshore Petroleum Board, which is part of the NEB, the company must provide a report about the reserve with reliable information and theories. The information must be scientifically valid without speculation to allow discussions between the company and the Board, who will decide whether the statement of discovery will be granted or not.

The Canadian legislation uses the National Instrument 51–101 Standards of Disclosure for Oil and Gas Activities, based on the COGEH definitions, for the classification and valuation of oil reserves.

COGEH is a set of three publications, which are interlinked, but with their own specific roles:

1. The first edition of Volume 1 (COGEH, 2002) was published by the SPEE to be a standard for the evaluation of oil and gas resources, and to establish the business protocols to properly prepare resource estimates. COGEH covers good business practices, qualifications of reserve auditors and includes definitions of oil and gas reserves and resources. In 2007, Volume 1 was updated and fully adopted the PRMS classification system shown in Fig. A4.1.
2. Subsequently, an expanded set of guidelines and examples was published in Volume 2 (COGEH, 2005), which was updated in 2014 to more clearly describe the treatment of contingent and prospective resources, known as "Resources Other Than Reserves" (ROTR).
3. Volume 3 was then published to discuss the evaluation of coalbed methane (CBM) accumulations, international properties, and bitumen reservoirs (COGEH, 2007).

While keeping elements of the principles basis of PRMS, the COGEH is more prescriptive in its approach, is linked to the Canadian regulatory disclosures, and offers detailed guidance on how it should be applied. This makes COGEH more appropriate for use by the stock market and auditors than PRMS. An updated version of COGEH was released in 2018.

### A4.3.1 Differences between COGEH and PRMS

As one would expect, by sharing the same heritage, and from the subsequent alignment of COGEH to PRMS, there are many similarities between the two systems. Yet, there are key differences, notably in the treatment of fuel gas, by-products, and "split conditions," with some being relatively more influential on the reserves booking.

- There is a common principle that all lease fuel is included in shrinkage, but PRMS gives the option to claim lease fuel as reserves, if it is offset by an operating expense and if such booking is not disallowed by the country's regulator. Many analysts disagree with booking fuel gas as reserves, as it is consumed in operations, and may overstate gas reserves, which are based on net sales volumes at a Reference Point of Sale.
- It is generally held that non-hydrocarbons that are removed prior to sales should not be included in reserves, yet COGEH permits associated sulfur to be reported separately as a reserve. This exception has arisen to address the needs of the significant heavy oil industry in Canada. While PRMS disallows non-hydrocarbon products to be booked as reserves, it does permit the revenue generated by their sale to be included in the economic evaluation of a project.
- COGEH applies different conditions for unproved reserves, whereas PRMS uses the same conditions for proved and unproved reserves. Most analysts agree with the PRMS approach, which works on the premise that the technical uncertainty in recovery from a project should be evaluated separately from its commercial maturity.

As mentioned above, a merger of PRMS and COGEH was agreed in principle, in 2011, with the hope of publishing a common global standard in 2016 (Etherington, 2014). Unfortunately, this has not been achieved and the key differences between the two systems remain unresolved.

## A4.4 NPD

In 1996, the NPD created a classification system that uses project maturity as a basis for classification, which allows linking of production, costs, and environmental data, and provides flexibility. Similar to a regulator, the NPD system maintains an associated document called, "Guidelines to classification of the petroleum resources on the Norwegian Continental Shelf (NCS)" that defines the principles for reserves classification in Norway. The NPD states that, *"Reserves comprise marketable oil resources, recoverable and remaining in an oil deposit, already licensed and approved for development or that has an exemption for the demonstration of the development plan."*

In 2001, the NPD system further developed the project maturity principle and became closely aligned with the earlier SPE-2001 system, which was replaced in 2007 by the PRMS. Project maturity is the explicit basis for the classification in PRMS, so the similarity with the NPD system is strong. The system is designed to ensure the Norwegian authorities obtain consistent reporting from licensees to allow accurate annual updates of the accounts for recoverable resources. In 2016, the current NPD system underwent some minor revisions to resource class divisions. Unlike the other systems, the current NPD guidance is mercifully brief at only 13 pages in length.

The NPD system uses four decision gates to describe the maturation of a project, and these milestones are partly incorporated as terms and conditions in production licenses on the NCS:

1. **Decide to initiate project**—Operator starts feasibility studies.
2. **Decide to concretize**—Operator identifies at least one technically and financially feasible concept that provides a basis for commencing studies that lead to concept selection.
3. **Decide to continue**—Operator selects a concept and decides to initiate further studies that lead to a decision to implement.
4. **Decide to implement**—Operator makes a final investment decision (FID) which results in the submission of an approved Plan for Development and Operation (PDO) or Plan for Installation and Operation (PIO).

The outcome could be a decision to take the project to the next phase, but it may be decided to postpone the project, or possibly to restart the work with a different set of preconditions. In all instances, the classification will reflect relevant project maturation. Companies cannot book reserves until after successfully passing through all the decision gates. Fig. A4.2 shows the connection between project maturation and resource classes, with corresponding activity in the stage.

One of the requirements of NPD system is the classifying and tracking the maturation of increased recovery projects. Some resource classes, as shown in Table A4.2, can have two attributes, F for (First) and A for

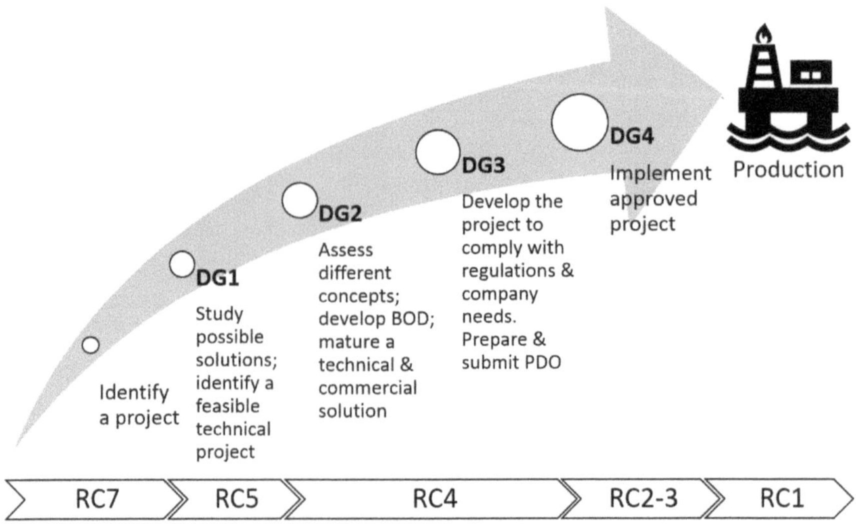


**Figure A4.2** Connection between decision gates, project maturation, and resource classes. *After NPD (2016b).*

(Additional) recoverable quantities from the same original oil and gas in place. Project category F denotes it is the (First) development project for one or more deposits, and the project must have a PDO or PDO exemption agreed with the NPD.

Project category A denotes when (Additional) recoverable petroleum volumes associated with the project lead to improved recovery of

**Table A4.2** Overview of resource classes, project categories, and uncertainty categories.

| Class | Resource class (sub-class) | Resource class | Project category | Uncertainty category |
|---|---|---|---|---|
| | Produced | RC0 | | |
| Reserves | In production | RC1 | | L, B, H |
| | Approved for production | RC2 | F, A | L, B, H |
| | Decided for production | RC3 | F,A | L, B, H |
| Contingent resources | Production in clarification phase | RC4 | F,A | L, B, H |
| | Production likely, but not clarified | RC5 | F, A | L, B, H |
| | Production unlikely | RC6 | | L, B, H |
| | Production not evaluated | RC7 | F, A | L, B, H |
| Undiscovered resources | Prospects | RC8 | | L, B, H |
| | Unmapped resources | RC9 | | L, B, H |

After NPD (2016b).

petroleum in place in deposits that are in production or projects classed as F. The uncertainty category is not used to define classes, but to indicate the Low, Base, High (L, B, H) uncertainty in the estimates.

## A4.4.1 Differences between NPD and PRMS

The NPD system is a brief, but easy to understand resource classification system, and differs from PRMS by acknowledging that several projects may be recovering oil and gas from the same accumulation, but they may be in different stages of maturity, and so they can be in different classes. The projects are distinguished from each other using the attributes, first project (F) and additional projects (A). Incremental recovery is tracked via the (A) attribute, and the quantities associated with primary recovery project use the (F) modifier, while the estimate of PIIP remains constant.

The NPD system is more focused on resource management with greater emphasis on capturing changes in EUR values than on yearly financial reporting. Hence, there is a regulatory directive for producing companies in Norway to provide detailed historical data and forecasts of production, costs, and investment, on an annual basis.

Scenario endpoints are similar to those in PRMS, with the low estimate being greater than P90 and high estimate being at least P10, but there is no differentiation between the endpoint deterministic and probabilistic estimates, which implies that the estimates should be the same irrespective of what method is used. Rather than employing a best estimate, the NPD system uses the base estimate, which reflects the current understanding of the extent, characteristics, and recovery factor of the reservoir. The base estimate can be calculated deterministically or stochastically and corresponds to the mean value (not the median or P50).

The concept of proved reserves or 1P, as defined by PRMS, is not recognized. Instead, P90 reserves are used as a simple, well-defined substitute, remembering that future, uncommitted projects are not allowed to contribute to either 2P or 3P reserves, as this would distort the P90 of the distribution. The terms Proved, Probable, and Possible are not utilized in the NPD system, but its definitions of low/1P, base/2P, and high/3P estimated quantities allow derivation of these entities if required.

Definitions of commercial/economic or subcommercial/subeconomic are not given by the NPD, which relies on the status categories to segregate reserves from contingent resources. The latter are defined as petroleum resources that have been found, but no decision has been made on their development. RC3 reserves are those where the licensees have yet to decide to recover and include projects for which the regulator has not yet approved a PDO or granted exemption. The differentiation of reserves from contingent resources seems to rest solely on the licensees' internal commitment to proceed with development. Under Norwegian petroleum law, licensees in the NCS are given the right to produce the petroleum, and approval of the PDO is an occasion to align interests to encourage the development to happen.

## A4.5 SEC

This rules-based US system, which is regulated by the SEC, is the oldest of the four systems with a definition of proved reserves being issued by the US regulator in 1978. The SEC administers federal securities laws, issues regulations to provide protection for investors, and ensures the markets are fair and honest. By insisting on full and effective disclosure to the investing public, the SEC provides a homogeneous base to compare companies, and so protects shareholders. However, it neither guarantees the return on an investment nor can it prevent the sale of securities that are of questionable value, so investors must make their own ultimate judgments of the worth of any securities on offer.

Regarding SEC documentation, revised regulations came into force on January 1, 2010, in a 160-page tome entitled "Modernization of Oil and Gas Reporting," in which reserves definitions are covered by regulation S-X, rule 210.4−10 (12 pages); regulation S-K Section 154 Items 1201−1208 cover disclosure requirements (17 pages). Another document, "Compliance and Disclosure Interpretations" (C&DIs), supplements the oil and gas rules and this was last updated on May 16, 2013. The SEC insists that annual 10-K filings by oil and gas corporations comply with these regulations and provides publicly available Comment Letters on individual company compliance (Taylor, 2017).

### A4.5.1 Differences between SEC and PRMS

The SEC only considers proved, probable, and possible reserves, and the reporting of 2P and 3P reserves is optional. Contingent and prospective resources are not permitted to be reported in the annual filing. Where analogs are used to estimate reserves, the assumed analogous reservoir properties must be, on aggregate, no better than those of the reservoir of interest.

With regard to development intent, schedule, and project finance, the SEC classifies Proven Undeveloped (PUD) locations as having undeveloped reserves if a development plan has been written and approved. Confirming evidence that the plan has got final investment decision (FID) approval must be provided. Thus, the PRMS reserves sub-class, "Justified for Development", does not exist under SEC rules, as all approvals need to be in place. The SEC insists that proved PUD locations must be drilled within 5 years and they must be debooked if not drilled. Under its definition of reserves, there must be, or at least a reasonable expectation that there exists all the necessary financing required to implement the project.

Forecasts of petroleum prices and development and operating costs do not include escalations to account for future conditions (unless prices are contractually defined, such as with long-term gas sales agreements). The flat forecast price is the average of reported prices on the first day of the month during the 12-month period prior to the "end date" stated in the report. Hence, 1P estimates of reserves using the SEC approach are invariably lower than those from the systems which allow petroleum prices to rise over time. The SEC bases its reserves on "economic producibility" rather than "commerciality," meaning the project is deemed to be economic if the computed NPV is positive at zero discount rate, i.e., if the project makes $1 profit it is "economic." The SEC regards commerciality as subjective, as it is linked to the hurdle rate of return required by a company before it will commit to a project. As this rate of return is likely to vary among companies, the SEC believes it reduces the comparability between disclosures. Yet, it is very unlikely that a potential buyer would ever consider investing in a project that is only economically viable at a zero-discount rate.

The SEC is somewhat confused over whether fuel gas should be booked as reserves. According to its documentation (see regulation S-K instruction 2 to item 1204) gas consumed in operations should be omitted from production data, but then (see regulation S-X rule 4—10) fails to clarify whether gas consumed in operations can be included as reserves. Consequently, 10-K filings by companies continue to include fuel gas as reserves (Taylor, 2017).

The SEC regulations do not allow non-hydrocarbon reserves or revenues obtained from their sale to be included in annual filings.

The upshot of these differences is that PRMS reserves figures for specific companies are invariably higher than SEC estimates for the same assets. Lee (2018) demonstrated that PRMS oil figures for major international oil and gas companies were higher; CNOOC by 101%; Petrobras by 28%; Rosneft by 18%; and Gazprom by 17%. The PRMS-SEC differences were even larger when considering natural gas, with Petrobras, Rosneft, and Gazprom PRMS estimates being 37%, 33%, and 23% higher, respectively. The main reasons for the differences are the debooking of projects under a constant price case and the strict enforcement of the 5-year rule by the SEC compared to the loosely constrained time to develop and the permitted use of an escalating price deck as defined in the PRMS.

In summary, the four resource classification regulations and guidelines described above are closely aligned, with the systems having the same overall intent. All the systems are maintained evergreen by independent bodies, who issue updates to keep the systems in line with best practices. PRMS is more suited for internal asset management, but being a public document, it can be tailored to users' needs. It offers some guidance on its application and is a popular system worldwide. The more prescriptive style of COGEH is better suited to regulatory and stock market reporting, and it has detailed guidance on how it should be applied, but the split conditions issue may blur the boundary between unproved reserves and contingent resources. The NPD system was created to provide the Norwegian Government with a full resource base of the country, both discovered and undiscovered, with corresponding reliable production forecasts, investments, and costs estimates. The system gives insight on environmental aspects of petroleum activity in Norwegian waters and provides input to the national policies on energy, finance, environment, and employment. It is used primarily by the Oslo stock market. The rules-based SEC is the most prescriptive and rigid of the four systems, being developed specifically for the US stock market, but is only concerned with proved and probable reserves. Although these systems are regarded as international standards and are used by various stock exchanges around the world, none are perfect.

# References


Aguilera, R., 1995. Naturally Fractured Reservoirs, second ed. Pennwell.

Alexander, J., Lohr, J., 1998. In: Risk analysis: lessons learned. Paper SPE-49030.

Alvaang, L., 2016. QC of Geomodels, Interoffice Correspondence with Lena Alvaang, Principal Geomodeller. Lloyd's Register, London.

Alvarado, V., et al., 2008. Screening strategy for chemical enhanced oil recovery in Wyoming basins, paper SPE-115940-MS. In: Annual Technical Conference and Exhibition, Denver, 21 September 2008.

Anderson, A., 2016. Analyst's Definition of 2P Reserves, Intercompany Correspondence With Alasdair Anderson, Principal Analyst. Wood Mackenzie, Edinburgh.

API, 2011. API RP 17A, Design and Operation of Subsea Production Systems - General Requirements and Recommendations, fourth ed. Global Standards. reaffirmed 2011.

Archie, G.E., 1942. The electrical resistivity log as an aid in determining some reservoir characteristics. Petroleum Transactions of AIME 146, 54–62.

Barry, R., 1993. The Management of International Oil Operations. Pennwell.

Beliveau, D., 2002. Blackjack, Heterogeneity & Horizontal Wells, Shell Canada, AAPG Distinguished Lecturer 2002–3.

Braaten, K., 2015. Understanding contingent resources. In: Presentation, REU Event, Houston.

Brown, A., February 2003. Improved interpretation of wireline pressure data. AAPG Bulletin 87 (2), 295–311.

Brown, M., Fosvold, L., Garza, J., Cook, D., 2000. A look to the past to avoid old traps in the future, improving the exploration process by learning from the past. In: Proceedings of Norwegian Petroleum Society Conference, Haugesund, Pp. 9–14, September 1998.

Câmara, R.J.B., Carvalho, L.F.J., Câmara, G.A.B., Rocha, P.S.M., 2015. Reserve certification: the actual scenario of the largest world oil producer countries and Colombia, paper OTC-26174-M, presented at Offshore Technology Conference Brasil, Rio de Janeiro, pp. 27–29. October 2015.

Capen, E.C., 1976. The difficulty in assessing uncertainty. Journal of Petroleum Technology 8, 843–850.

Carroll, L., 1871. Through the looking-glass, and what alice found there. Classic Novel.

COGEH, 2002. Canadian Handbook of Oil & Gas Evaluation Handbook. In: Reserves Definitions and Evaluation Practices and Procedures, Aligned with SPE PRMS and Updated in 2007, vol. 1.

COGEH, 2005. Canadian Handbook of Oil & Gas Evaluation Handbook. In: Detailed Guidelines for Estimation and Classification of Oil and Gas Resources and Reserves, Updated for Resources Other Than Reserves (ROTR) in July 2014, vol. 2.

COGEH, 2007. Canadian Handbook of Oil & Gas Evaluation Handbook. In: CBM Reserves and Resources/International Properties/Bitumen and SAGD Reserves Resources, Published in September 2007, Updated Section on Bitumen and SAGD Reserves and Resources in April 2014, vol. 3.

Collarini, 2000. True Vertical Depth Correction, Online Presentation by Collarini Engineering Inc.. http://www.collariniexperts.com/assets/tvt_presentation.pdf.

Craig, F.F., 1971. The reservoir engineering aspects of waterflooding. SPE Monograph 2, 64–66.

Cronquist, C., 1991. Reserves and probabilities: synergism or anachronism? Paper 23586-PA. SPE Journal.

Dean, E.W., Stark, D.D., 1920. A convenient method for the determination of water in petroleum and other organic emulsions. Journal of Industrial and Engineering Chemistry 12, 486–490.

Deloitte, 2013. Reserves matter(s) – E&P Roundtable. Oil & Gas Spotlight (1).
Demirmen, F., 2007. Reserves estimation: the challenge for the industry. Journal of Petroleum Technology 5, 80–89.
DPA/SIPES, 1991. Confidentiality agreement. In: Division of Professional Affairs of the American Association of Petroleum Geologists and Society of Independent Professional Earth Scientists, Document Template.
Dykstra, H., Parsons, R., 1950. The prediction of oil recovery by water flood. In: Secondary Recovery of Oil in the United States, second ed. American Petroleum Institute.
Etherington, J., 2014. Merging PRMS and COGEH, a status update. In: Presentation at Ryder Scott Canada Reserves Conference, Calgary, 2 October 2014.
Feynman, R., 1985. Letter to Armando Garcia J, 11 December 1985.
Harrell, R., Hodgin, J., Wagenhofer, T., 2004. Oil and gas reserves estimates: recurring mistakes and errors. In: Paper 91069, SPE Annual Technical Conference and Exhibition, Houston.
Harper, F., 2000. Prediction accuracy in petroleum prospect assessment: a 15 year retrospective in BP. In: Proceedings of Norwegian Petroleum Society Conference, Improving the Exploration Process by Learning from the Past, vol. 9, pp. 15–21.
Harrison, R., 2012. Framework increases effectiveness of M&A teams when acquiring information from data rooms. In: Paper 159360, SPE Annual Technical Conference & Exhibition, Houston.
Harrison, R., 2013. Putting a value on estimated NGL volumes in prospect evaluation. In: Paper 166409, SPE Annual Technical Conference & Exhibition, New Orleans.
Harrison, R., 2014a. Reading between the lines of A&D documentation. In: Paper 169870, SPE Hydrocarbon Economics & Evaluation Symposium, Houston.
Harrison, R., Falcone, F., 2015. Re-grading proven reserves using PRMS. In: Paper 177244, SPE Latin American & Caribbean Petroleum Engineering Conference, Quito.
Harrison, R., Falcone, F., 2017. Handling decommissioning and restoration liabilities within PRMS and their impact on reported reserves of producing fields approaching abandonment. In: Paper 185497, SPE Latin American & Caribbean Petroleum Engineering Conference, Buenos Aries.
Harrison, R., Quinn, P., March 2018. Review suggests draft SPE PRMS 2017 unfit for public resource reporting. Oil & Gas Journal.
Hayes, M., March 2016. Why Can't I Get My Reserves Right? Dinner Meeting Presentation to SPE London Chapter.
Hattingh, S., 2008. Technical Due Diligence in the Upstream Petroleum Business. Gaffney Cline & Associates, Focus Newsletter (45).
Heavysege, R.G., 2002. Formation evaluation of fresh water shaly sands of the Malay basin, offshore Malaysia. In: SPWLA 43rd Annual Logging Symposium, June 2002.
Hodgin, J., Harrell, R., 2006. The selection, application, and misapplication of reservoir analogs for the estimation of petroleum reserves. In: Paper SPE-102505, Presented at the Annual Technical Conference and Exhibition, San Antonio, Texas, 24–27 September 2006.
HKeX, September 2009. Consultation Paper on New Listing Rules for Mineral and Exploration Companies. Hong Kong Exchange.
Isa, M.F.M., Azhar, M.A., 2009. Meeting Technical Challenges in Developing High $CO_2$. Gas Field Offshore. Petronas Carigali.
LaPlante, A., 2006. How Do You Successfully Merge Two Corporate Cultures? Stanford Business Insights.
Lea, J.F., Nickens, H.V., 2004. Solving gas well liquid-loading problems. Paper SPE-72092-JPT. SPE Journal of Petroleum Technology 56 (04).
Lee, J., 27 July 2018. Veteran expert says PRMS may be better reserves reporting method than SEC, talk at topical luncheon, UrTEC, Houston. World Oil.

Logicom, E.&P., 2018. "REP"; Resource Evaluation Program, Uses the Monte-Carlo Technique for a Rigorous Solution of the Volumetrics Equation. https://www.logicomep.com/products/rep/.

McCain, W.D., et al., 2011. Petroleum Reservoir Fluid Property Correlations. PennWell Publishing Co.

McLane, M., et al., 2008. Responsible reporting of uncertain petroleum reserves. AAPG Bulletin 92 (10), 1431–1452.

Morales, E., 2016. The importance of assumptions in the estimation, classification and categorization of reserves and resources. In: 6th Tunisia Oil & Gas Summit, 29 September 2016.

Norwegian Oil & Gas Association, 2011. Recommended Guidelines for Use of Confidentiality Agreements in Data Rooms. Guideline No. 57, Revision 2.

NOGA, 2011. Recommended guidelines for use of confidentiality agreements in data rooms. Norwegian Oil & Gas Association, Guideline No. 57, Revision 2.

NPD, 2016a. The Resource Report - Petroleum Resources on the Norwegian Continental Shelf.

NPD, August 2016b. The Norwegian Petroleum Directorate's Resource Classification System.

NPD, 2019. Recovery Factor. https://www.npd.no/en/facts/production/improved-oil-recovery-ior/recovery-factor/.

Offshore, 1 June 2011. TAQA Breathing New Life into Aging North Sea Platform. Offshore Magazine Article. Penwell.

Ofstad, K., Overetveit, A., Kullerund, L., Heggland, K., 2000a. Probability of discovery and the reasons for dry wells: results from the project (article 3). In: Improving the Exploration Process by Learning From the Past: Proceedings of Norwegian Petroleum Society Conference, Haugesund, Pp. 47–55, September 1998.

Ofstad, K., Kullerund, L., Hellisken, D., 2000b. Evaluation of Norwegian wildcat wells (article 1). In: Improving the Exploration Process by Learning From the Past: Proceedings of Norwegian Petroleum Society Conference, Haugesund, Pp. 15–21 September 1998.

Oil & Gas Authority, 2016. Website Gateway to the Release of UK Historical Production Data, Complete Production History for Oil Fields (From 1975) and Gas Fields (From 1995). https://itportal.decc.gov.uk/pprs/full_production.htm.

Oil & Gas Journal, 2004. Royal Dutch/Shell Lowers Proved Reserves Estimate 20%.

Olsen, G., Lee, J., Blasingame, T., 2011. Reserves overstatements – the problem we're finally going to talk about. In: Paper 134014, SPE Economics and Management, pp. 68–78.

Otis, R., Schneidermann, N., 1997. A process for evaluating exploration prospects. American Association of Petroleum Geologists Bulletin 81, 1087–1108.

Petroleum Solutions, 2018a. "RE Toolkit"; A General Purpose Reservoir Engineering Application. www.petroleumsolutions.co.uk/downloads.html.

Petroleum Solutions, 2018b. "Waterdrive"; A Collection of Classical Waterdrive Calculations and Routines. www.petroleumsolutions.co.uk/downloads.html.

PwC, 2012. Corporate divestitures from strategy to execution. Four guiding principles to optimize value. Price Waterhouse Coopers, Deals practice 2012.

PwC, 2013. Executing a successful listing: markets for oil & gas. IPO publication by Price Waterhouse Coopers. October 2013.

Quinn, P., 2016. Personal Correspondence on QC of Dynamic Models With Patrick Quinn, Reservoir Engineering Discipline Lead. Lloyd's Register, Aberdeen.

Ramachandran, V.S., Rogers-Ramachandran, D., 2010. Reading Between the Lines: How We See Hidden Objects. Scientific American.

Reeder, M., 2017. Oil & gas sector company listings and resource reporting systems used in the stock markets of Asia Pacific region. In: Interoffice Communication With Mike Reeder, LR Singapore, 26 May 2017.

Rose, P., 2001. Risk Analysis and Management of Petroleum Exploration Ventures: AAPG Methods in Exploration 12.

Rushworth, S., 2004. Going international: tools and techniques for assessing new country entry: UK North Sea example. In: Presentation by IHS Energy at CSPG, May 2004.

Scott, Ryder, 2018. In: Reservoir Solutions, quarterly newsletter 21 (3), 12–13. July-September 2018.

Scallon, J., 2015. Personal Correspondence on Prospective Resource Assessments With Jim Scallon, Principal Geologist. Lloyd's Register, London.

SCM, 2017. Proposed policy for the admission of mineral or oil and gas exploration or extraction corporations or assets to the main market of Bursa Malaysia Securities BHD. In: Response to Public Consultation Paper by Securities Commission Malaysia, 19 January 2017.

Schmalz, J., Rahme, H., 1950. The Variation of Waterflood Performance With Variation in Permeability Profile. Production Monthly.

Scott, M.J., 2012. Oil and gas reserves and resources reporting - the market rules. In: Paper 155398-MS, Presented at SPE Asia Pacific Oil & Gas Conference & Exhibition, Perth, October 2012.

Seager, R., 2013. An overview of the petroleum resources management system and its relationship to UNFC. In: OGRC Presentation to UNECE, Geneva, 23 April 2013.

SGX, 2013. Practice note 6.3 disclosure requirements for mineral, oil and gas companies. In: Mainboard Rules, Singapore Exchange, September 2013.

Shepherd, M., 2009. Oil Field Production Geology. AAPG Memoir 91.

SPE, 2018. Petroleum Resources Management System.

Stiles, W.E., 1949. Use of Permeability Distribution in Water Flood Calculations, vol. 186 (9). Trans AIME.

Stoneley, R., 1995. North Sea Petroleum Plays. Introduction to Petroleum Exploration for Non-geologists. Oxford University Press, p. 106.

Taylor, P., 2017. Reserves reporting – a comparison of the three major standards. In: Presentation at the SPEE Europe Chapter Meeting, London, 7 September 2017.

Tversky, A., Kahneman, D., 1974. Judgment under uncertainty: heuristics and biases. Science 185 (4157), 1124–1131.

USGS, 1976. Principles of the Mineral Resource Classification System for the U.S. Geological Survey. USGS Bulletin, 1450-A, Washington, DC.

Weber, K.J., Bakker, M., 1981. Fracture and vuggy porosity. In: Paper SPE-10332, Presented at the Annual Technical Conference and Exhibition, San Antonio, Texas, 4–7 October 1981.

Wright, S.G., 2009. The confidentiality agreement. PBLA Take-off 65 (54).

# Index

'*Note*: Page numbers followed by "t" indicate tables and "f" indicate figures.'

FSC
www.fsc.org
MIX
Papier | Fördert
gute Waldnutzung
FSC® C083411

Kolibri 360 GmbH
Ferdinand-Jühlke-Straße 7
99095 Erfurt, Deutschland
produktsicherheit@kolibri360.de